The Carbon Calculation

Critical Green Engagements

Investigating the Green Economy and Its Alternatives

James Igoe, Molly Doane, Tracey Heatherington, Melissa Checker, José E. Martinez-Reyes, and Mary Mostafanezhad
SERIES EDITORS

The Carbon Calculation

Global Climate Policy, Forests, and Transnational Governance in Brazil and Mozambique

Raquel Rodrigues Machaqueiro

THE UNIVERSITY OF ARIZONA PRESS
TUCSON

The University of Arizona Press
www.uapress.arizona.edu

We respectfully acknowledge the University of Arizona is on the land and territories of Indigenous peoples. Today, Arizona is home to twenty-two federally recognized tribes, with Tucson being home to the O'odham and the Yaqui. Committed to diversity and inclusion, the University strives to build sustainable relationships with sovereign Native Nations and Indigenous communities through education offerings, partnerships, and community service.

© 2023 by The Arizona Board of Regents
All rights reserved. Published 2023
First paperback edition published 2025

ISBN-13: 978-0-8165-4663-3 (hardcover)
ISBN-13: 978-0-8165-5658-8 (paperback)
ISBN-13: 978-0-8165-4664-0 (ebook)

Cover design by Leigh McDonald
Cover photos by Raquel Rodrigues Machaqueiro

Publication of this book is made possible in part by the proceeds of a permanent endowment created with the assistance of a Challenge Grant from the National Endowment for the Humanities, a federal agency.

Library of Congress Cataloging-in-Publication Data are available at the Library of Congress.

Printed in the United States of America
♾ This paper meets the requirements of ANSI/NISO Z39.48-1992 (Permanence of Paper).

Contents

List of Illustrations		*vii*
Acknowledgments		*ix*
Abbreviations		*xiii*
	Introduction	3
1.	Carbon and the Synecdochical Re-inscription of Science	35
2.	The UNFCCC Negotiation Process: Agreeing to Keep Talking	65
3.	Forests at the Center of Climate Change Policies	103
4.	Operating Transnational Governance Locally: REDD in Acre and Mozambique	138
5.	REDD on the Ground: Broadening the Scope, Fostering Vagueness and Confusion	188
	Conclusion	245
Notes		*255*
References		*279*
Index		*301*

Illustrations

Figures
1. Sentence formulations — 75
2. Recursivity of the UNFCCC's language — 78
3. Forest monitoring systems — 87
4. Drivers of deforestation — 91
5. Transparency, accuracy, consistency — 94
6. Road to Xapuri, once the capital of rubber production — 146
7. Forum on REDD held in Quelimane, Zambézia — 180
8. Cattle in a deforested area near Xapuri — 201

Maps
1. Map of Brazil / Acre — 32
2. Map of Mozambique / Zambézia — 33

Boxes
1. Chico Mendes: From union leader to environmentalist — 145
2. Civil war in Mozambique — 166
3. Land tenure: Colonial continuities — 171
4. Shifting agriculture — 181
5. Operation Amazonia: Developing the forest — 200
6. Rubber production in Acre — 204
7. From *cantineiros* to emerging farmers — 225
8. Planting eucalyptus to promote forest conservation — 235

Tables
1. UNFCCC Conferences of the Parties — 66
2. Annual variation of deforestation rates in Acre — 158

Acknowledgments

This work is indebted to many people across multiple places. I cannot name all of them, but I want to start by thanking all the people in Brazil, Mozambique, and Washington, D.C., who kindly agreed to spend time with me and answer all my questions. The beginning of my research in Washington, D.C., was facilitated by Mary and Chuck Kent, who generously provided a home, advice, and friendship throughout these years. Prior to the beginning of my doctoral program, I met Florence Daviet and Heather McGray, who hosted me at the World Resources Institute for a short internship, during which we sustained many productive discussions about REDD. Those discussions set the field for the beginning of this journey.

My research in Brazil would have not been possible without the friendship of Eduardo Caetano da Silva, who has been a friend and an intellectual partner since 2001. His family was very generous and provided me with a home in Campinas. Mauro de Almeida opened all the doors I needed in Brasília and in Acre. I am greatly indebted to friends Mariana Pantoja, Elder Andrade de Paula and Patricia, Michel Schmidlehner, and Dercy, who provided (in many different ways) enormous support. I am also thankful to Flávio, Marcelo Piedrafita Iglesias, Ariana, and Jade for their company, support, and meaningful conversations. The Federal University of Acre (UFAC) gave me an institutional affiliation and offered me the honor of giving two talks, during which I had the opportunity to discuss and share some preliminary results of this research.

In Mozambique, I am greatly indebted to the friendship and support given by Euclides Gonçalves, Décio Muianga, Joca, Anésio, and Jasmine. The time I spent with all of them eased the fieldwork (especially at times of frustration) and made me look at things in a new perspective. I also thank Francis Masse, Filipe Mate, and Rui Assubuji for their support and engaging conversations throughout fieldwork. António Sopa from the Historical Archive provided great help and guidance during archival research. The University Eduardo Mondlane granted me with a valuable academic affiliation (that opened many doors) and invited me to present part of my research.

In Portugal, I am deeply indebted to all the staff at the Arquivo Histórico Ultramarino (the historical archives). Their patience and commitment in fostering the best research environment possible (even in the face of crippling budget cuts and understaffing) is priceless. All of my conversations with Paulo Granjo have been inspiring and helped me understand Mozambique with different eyes. I also thank Manuela (who always had a home for me), Paula, and Zé for their endless friendship and support. My gratitude goes as well to Joana and Manuel for their beautiful friendship, sense of humor, and attentive reading.

My parents have been my greatest force. Their unconditional love and model of/for how to live life, made much of what I am today. Unfortunately, my father did not live long enough to see this new journey in my life, but I know he would be happy. My deepest gratitude goes to my mother who has always been there for me no matter what. In countless moments, Lenny made my life sunnier and happier (even if he is not always aware of that)—I thank him for that and for his infinite love. I am also grateful to my sister and brother (and their families), as well as my aunt Bia, for their unconditional support and cheerfulness.

I am deeply indebted to the Anthropology Department at the George Washington University for creating the best doctoral program ever, and for allowing me to be one of the first students to prove it. I thank my doctoral colleagues, especially Jessica, Chloe, and Evy, as well as Jonathan Higman for making the Hortense Amsterdam House such a special place. I would also like to acknowledge the mentoring given by Sarah Wagner, Richard Grinker, Joel Kuipers, Bob Shepherd, Joshua Bell, and David Braun, who helped me throughout my doctoral program at GWU.

Acknowledgments

The members of my dissertation committee deserve my greatest appreciation for their guidance, support, and commitment to my success. This work would not have been possible without the friendship and demanding (but always constructive) criticism of Steve Lubkemann, Alex Dent, Ilana Feldman, and Hugh Gusterson. Edward LiPuma joined this work at a later stage but was a source of inspiration very early on.

Finally, I want to express my gratitude toward Sofia Lovegrove, who helped me cut unnecessary words, as well the editors of University of Arizona Press, Allyson Carter and Melissa Checker, for their support and for providing two anonymous reviewers who pushed me to clarify my arguments and make this a better book. The limitations of this work are, of course, all my own.

I acknowledge the financial support for this research since its inception from the Lewis N. Cotlow Foundation and the Institute for Ethnographic Research at GWU, and the Portuguese Science and Technology Foundation, which provided me with a generous grant that allowed me to finish my doctoral program and write without financial concerns—POCH (SFRH/BD/117856/2016).

Abbreviations

BNDES	Banco Nacional de Desenvolvimento Econômico e Social (Brazilian National Development Bank)
CARE	Cooperative for Assistance and Relief Everywhere
CCBA	Climate Community and Biodiversity Alliance
CCBS	Climate Community and Biodiversity Standards
CDM	Clean Development Mechanism
CDSA	Companhia de Desenvolvimento de Serviços Ambientais (Company for the Development of Environmental Services)
CER	Certified Emissions Reductions
CEVA	Comissão Estadual de Validação e Acompanhamento (State Commission of Validation and Monitoring)
CFCs	Chlorofluorocarbons
CfRN	Coalition for Rainforest Nations
CIFOR	Center for International Forestry Research
CIMI	Conselho Indigenista Missionário (Indigenist Missionary Council)
COP	Conference of the Parties
CPLC	Carbon Pricing Leadership Coalition
CTV	Centro Terra Viva

DUAT	Direito ao Uso e Aproveitamento da Terra (Right to Use and Harness the Land)
EDF	Environmental Defense Fund
EMBRAPA	Empresa Brasileira de Pesquisa Agro-Pecuária (Brazilian Agricultural Research Corporation)
ETS	Emissions Trading Scheme
FAO	Food and Agriculture Organization
FAS	Fundação Amazonas Sustentável (Amazonas Sustainable Foundation)
FCPF	Forest Carbon Partnership Facility
FERN	Forests and the European Union Resource Network
FGV	Fundação Getúlio Vargas (Getúlio Vargas Foundation)
FIP	Forest Investment Plan
FRELIMO	Frente de Libertação de Moçambique (Mozambique Liberation Front)
FUNAI	Fundação Nacional do Índio (National Indian Foundation)
GCF	Governors' Climate and Forests Task Force
GHGs	greenhouse gases
GIZ	Deutsche Gesellschaft für Internationale Zusammenarbeit (German Corporation for International Cooperation)
IBAMA	Instituto Brasileiro do Meio Ambiente e dos Recursos Naturais Renováveis (Brazilian Institute of Environment and Renewable Natural Resources)
ICAP	International Carbon Action Partnership
IET	International Emissions Trading
IFC	International Finance Corporation
IIAM	Instituto de Investigação Agrária de Moçambique (Agricultural Research Institute of Mozambique)
IIED	International Institute for Environment and Development
IMC	Instituto de Mudanças Climáticas (Institute of Climate Change)
INDC	Intended Nationally Determined Contribution
IPAM	Instituto de Pesquisa Ambiental da Amazônia (Institute for Environmental Research in Amazonia)
IPCC	Intergovernmental Panel on Climate Change
ITERACRE	Instituto de Terras do Acre (Acre's Land Institute)
ITMO	Internationally Transferred Mitigation Outcomes
IUCN	International Union for Conservation of Nature

JA!	Justiça Ambiental (Environmental Justice)
JI	Joint Implementation
JICA	Japan International Cooperation Agency
JNR	Jurisdictional Nested REDD
KfW	Kreditanstalt für Wiederaufbau (German Investment and Development Bank)
MASA	Ministério da Agricultura e Segurança Ambiental (Ministry of Agriculture and Food Security Mozambique)
MICOA	Ministério da Coordenação Ambiental (Ministry of Coordination of Environmental Affairs—Mozambique)
MINAG	Ministério da Agricultura (Ministry of Agriculture—Mozambique)
MITADER	Ministério da Terra, Ambiente e Desenvolvimento Rural (Ministry of Land, Environment, and Rural Development—Mozambique)
MRV	measuring, reporting, and verifying
MST	Movimento dos Trabalhadores Rurais Sem Terra (landless workers' movement)
NCBs	non-carbon benefits
NDC	nationally determined contribution
NGO	nongovernmental organization
OECD	Organization for Economic Co-operation and Development
PGTA	Plano de Gestão Territorial e Ambiental (Territorial and Environmental Management Plan)
PPG7	pilot program to conserve the Brazilian rainforest (financed with grants from the G7)
ppm	parts per million
REDD(+)	Reduced Emissions from Deforestation and forest Degradation
REM	REDD for Early Movers
RENAMO	Resistência Nacional Moçambicana (Mozambican National Resistance)
R-PIN	readiness plan idea note
R-PP	readiness preparation proposal
SEDENS	Secretaria de Estado de Desenvolvimento Sustentável (State Secretariat of Forest Development, Industry, Commerce, and Sustainable Services)

SISA	Sistema de Incentivos a Serviços Ambientais (System of Incentives for Environmental Services)
tCER	temporary Certified Emission Reductions
tCO2e	ton of carbon dioxide equivalent
TEEB	The Economics of Ecosystem and Biodiversity
TINA	there is no alternative
TIST	The International Small Group & Tree Planting Program
TNC	The Nature Conservancy
UFAC	Universidade Federal do Acre (Federal University of Acre)
UNDP	United Nations Development Program
UNEP	United Nations Environmental Program
UNFCCC	United Nations Framework Convention on Climate Change
USAID	United States Agency for International Development
UT-REDD	Unidade Técnica do REDD (REDD technical unit)
VCS	Verified Carbon Standard
WMO	World Meteorological Organization
WRM	World Rainforest Movement
WWF	World Wildlife Fund

The Carbon Calculation

Introduction

Considering Responsibility: Paris in Maputo

In March 2016, the Mozambican government organized a public conference to divulge the outcomes of the Paris Conference, celebrated in December of the previous year. Despite the little information available about the event, the conference room at the four-star hotel gathered around three hundred people, including government officials, practitioners from the numerous NGOs and development agencies that populate Maputo's landscape, business representatives, and university students. The atmosphere was very formal, moderated by a well-known reporter from the major private TV channel, with the minister of the environment, Celso Correia, expected to open the event. After a long period of waiting, the audience was finally told that the minister was about to enter the room. When the audience was finally silent and standing up, the minister entered the room and was greeted with applause.

His intervention was short but significant. After stating that the Paris Agreement was not satisfactory—given its lack of ambition—the minister emphasized that the country had to face the reality of climate change, which he deemed responsible for floods in the north and droughts in the center and south, and to which he attributed incalculable damages and losses. Accordingly, Mozambique was a signatory to the agreement and had to comply with it, assuming its own responsibilities. For the minister, assuming responsibilities to reduce emissions (referred to as mitigation) would not be hard for a country like Mozambique because, he opined, taking care of its natural resources and biodiversity had always been part of Mozambique's culture.

Finally, he stated that the biggest mitigation contribution of the country was the wide-ranging forest reform launched recently by the new government, which was intended to protect Mozambique's miombo forests.

The main message stated by the minister—that *Mozambique has to assume its mitigation responsibilities*—was repeated in different forms by the speakers who followed, either Mozambican officials, or international NGO representatives. This apparent consensus on responsibility was, however, ruptured by the representative of a group of local NGOs. This activist not only bemoaned the Paris Agreement for its weak and vague targets, but also for its lack of any provisions for addressing climate change impacts and helping the most vulnerable countries—Mozambique being one of them. Concluding his intervention by calling for a radical change of the current economic model, he deplored the hegemony of market-based solutions within the United Nations Framework Convention for Climate Change (UNFCCC), and the lack of differentiation between the levels of responsibility demanded from each country.

But the most surprising moment of the event came later, when the debate opened to the audience. While the message of the last speaker—about the overall unfairness of the Paris Agreement—was left unanswered, a couple of Mozambican students from the university proposed a new meaning for "responsibility," rejecting the position of Mozambique as a "developing" country within the climate negotiations. Instead of placing Mozambique in the group of countries waiting for the decisions and financial aid from other countries, these students called for the government to implement its own strategies to address climate change, including the creation of a national fund to cope with damages and compensations in the face of extreme climate events. Despite this assertion of sovereignty, the students reflected national anxieties about Mozambique's poverty levels and dependence on foreign aid. These suggestions simultaneously expressed principles of individualism, self-government, and autonomy that are, overall, constitutive of a neoliberal mode of governance (Agrawal 2005; West 2008). Neoliberalism, a prescriptive concept regulating the relationship between the state, capital, property, and individuals (Ganti 2014) is characterized by an overreliance on market-based solutions (against state-based regulations and policies) followed by an increasing deregulation and reregulation (Harvey 2005; Castree 2008), the prevalence of the individual over the collective (expressed in the value of individual freedom), and the decentralization of spaces of decision to NGOs and private companies (Fletcher 2010; Fletcher and Büscher 2017)—all of which compounding concomitant subjectivities aligned with values of individualism and market competition (Foucault 1991; Ganti 2014).

This neoliberal mode of governance—especially in its transnational iterations—is the main object of this book. More specifically, I examine how climate governance has become increasingly entangled with neoliberal principles, and how this neoliberalization has important implications at the political level (the responsibility for climate change has gradually shifted from corporations in the industrialized worlds into an unspecified mass of individuals); at the policy level (measures to solve climate change have mostly relied on market mechanisms); and at the environmental level. This event in Maputo provides a good example of how transnational organizations (like the UNFCCC) are able to gradually instill new perspectives over the world and its problems; in this case, how subjects are "increasingly 'empowered' to discipline themselves" to be *responsible* (Ferguson and Gupta 2002, 989). As a result, Mozambican students expressed a notion of responsibility that not only ignores the history of the problem of climate change and the obligations of industrialized countries toward poor ones, but also adheres to neoliberal principles of self-government and autonomy.[1] The ingraining of these neoliberal perspectives over climate change—to the extent that solutions to climate change are always discussed within the same economic model that created the problem in the first place—has been a very gradual (and contested) process, but ultimately consolidated in the Paris Accord.

Paris 2015: A Landmark Agreement in the History of Negotiations

Considered the most important agreement in the history of climate change conferences, the Paris Accord was celebrated in 2015 and presented to the public as a successful milestone, tantamount to the one achieved with the Kyoto Protocol in 1997. For the UNFCCC, Paris was a success for bringing together more countries than any other previous agreement, and for establishing a new form of addressing the problem since "for the first time [the Paris Agreement] brings all nations into a common cause to undertake ambitious efforts to combat climate change and adapt to its effect, with enhanced support to assist developing countries to do so. As such, it charts a new course in the global climate effort" (UNFCCC 2017).

The agreement was achieved on December 12, 2015, and by November 2016, countries representing 55 percent of global emissions had already ratified it, followed by another 127 countries by January 2017. The fact that the Paris Agreement is based on voluntary contributions and not on legally defined emissions reductions has probably facilitated the ratification process. Indeed, the voluntary character of these mitigation efforts is enshrined in the language of the accord, which merely *encourages* developing parties "to move toward economy-wide targets over time in the light of different national circumstances" (UNFCCC 2017), thus replacing the once legally binding commitments defined in Kyoto. Despite the fact that the Kyoto commitments were already weak—given that there was never a mechanism to assess parties' achievements or consequences for noncompliant parties—the Paris Agreement only reinforced this status quo, since it "includes a mechanism that will facilitate implementation and promote compliance in a non-adversarial and non-punitive manner" (UNFCCC 2017).

The success of the Paris Agreement has been, however, relativized by some environmentalists and activists who consider it "woefully inadequate" (Klein 2016) given its incapacity to ensure that temperatures do not rise more than the two degrees Celsius, not to mention the absence of any concrete measures to deal with climate change impacts, or clear references to actual sources of financing for poor countries. The agreement is, however, a paradigmatic example of how the UNFCCC has been incredibly successful at keeping parties talking about the problem of climate change:

> [The] plan is that governments *agree to meet every five years* and hope that things have changed for the better. (Klein 2016)[2]

But the terms of the talk have gradually been changing. The accord represents, indeed, a pivotal change in the course of the climate negotiations—namely, in the kind of engagement by the parties, on the types of parties involved, and in the temporal stance adopted by the parties.

A New Direction in Climate Governance

If negotiations that have been ongoing since 1992, without tangible results, can be easily conceived of as an utter failure—especially in the face of increasing emission levels—the process led by the UNFCCC has to be acknowledged instead, as a significant success, albeit of a particular sort and for particular parties. Continuing to talk throughout all these years is a successful outcome of these negotiations—not just because they produced a narrative enabling the enrollment of all parties (Mosse 2005) and the legitimacy of the UNFCCC as its governing body, but also because such enrollment has facilitated the increased involvement of poor countries in the solutions to curtail emissions (see chapter 2). While parties continue to talk, the conversation has gradually changed from one in which the responsibility of industrial countries for the state of the climate was assumed to a situation in which that responsibility has been subtly redistributed, so that poor countries are now deemed equally responsible for acting against climate change. As I witnessed in Mozambique, this responsibility has apparently come to be accepted as an imperative. My discussion of the history of these negotiations will demonstrate how countries from the Global North were able to gradually change the focus of the responsibilities regarding climate change from a historical perspective into an orientation toward the future (see chapter 3).

The new direction in climate governance established by Paris embodies this shift in responsibility as well as two other important features: an overall acceptance of the central role of market mechanisms in reducing emissions and a decentralization of actions with non-country parties taking on an increasing role in mitigation initiatives. The Kyoto Protocol had defined market instruments (called flexible mechanisms or carbon markets) as an aid to countries in their efforts to achieve their emissions-reduction goals. Although carbon markets have struggled to succeed, their performative failures and repeated crises have not led to their extinction. Rather, the Paris Agreement expanded market initiatives. In the first chapter, I explain how and why these markets have persisted and are being expanded despite their obvious under-performance.

The Paris Agreement also decentralized these markets, which means that instead of reforming the market instruments already in place, parties decided to open them to private stakeholders and individual initiatives who are now allowed to operate independently and outside of UNFCCC's oversight. The call for "all non-Party stakeholders to address and respond to climate change, including those of civil society, the private sector, financial institutions, cities and other sub-national authorities" (UNFCCC 2017) provides evidence that the spirit of Kyoto, of *common but different responsibilities*, has been lost. Moreover, the notion of responsibility itself is undergoing further discursive redistribution, shifting weight from the industrialized world to the Global South, and from states to individuals, or non-state actors. By diverting attention to individuals' actions—namely consumption choices—this new type of governance depoliticizes the problem of climate change (and its origins) while also undermining the institutional response that it requires (Maniates 2001). In sum, the current climate governance is eminently depoliticized, dehistoricized, decentralized, and informed by a neoliberal perspective that favors market approaches and the action of private actors. This has important implications.

The central idea undergirding this book is that the current climate governance offers relevant insights into what I consider to be new modalities of transnational governance (be them over the climate or the environment, democracy, or the economy). My foundational argument is, thus, that the particular ways in which the problem of climate change has been defined have enabled the emergence of new and powerful forms of neoliberal transnational governance. Before going any further, I will clarify what I mean by transnational governance.

Transnational Governance and Political Agency

Following Ferguson and Gupta, I understand transnational governance as a form of global governance that moves across various jurisdictional levels—international, national, and local—that depends on the mobility of people, objects, and ideas and that seeks to further the strategic interests of certain states or, more importantly, economic groups, by deploying common spaces and/or values—such as the global planet, or environmental conservation (2002, 996). Transnational governance operates through actions and ideas that "may be embedded in the daily practices of nation-states" or that "may crosscut or superimpose themselves on the territorial jurisdiction of nation-states" (996)—but they always depend on the collaboration (whether intended or not) of the state authorities upon which that governance will act (whether benefiting those states, other states, or other stakeholders).

I will thus be examining transnational governance in the context of climate change policymaking—how climate change policies are negotiated and created at an international level; how different states exercise their political agency in these negotiations; how these policies are then nationally implemented and locally negotiated and reinterpreted; and how the circulation of these policies is continuously put in motion in different directions and across multiple jurisdictional levels. The emergence of climate change as the biggest environmental crisis ever faced by the planet has brought to the fore, in very compelling ways, the contradictions between environmental protection and the market logics of capitalism (Klein 2014), while exacerbating many of the anxieties and contradictions regarding the political agency of national authorities in the context of international agreements that aim to tackle global problems like climate change.

If climate change is indeed the biggest challenge ever faced by humans, the creation and implementation of policies to tackle it should be an unchallengeable priority to ensure the survival of humanity and the planet as we know it, and yet this negotiation process has been rife with foot-dragging, conflict, and contradictions that not even the Kyoto Protocol solved. How are we to make sense of a negotiation process that has been going on for more than twenty years without successful results? How can we understand the several agreements achieved when they embody policies that neglect the historic responsibility of industrialized countries in the concentration of greenhouse gases (GHGs) in the Earth's atmosphere, and that put the heaviest load of mitigating these gases upon countries already burdened by poverty and lack of infrastructure; or provide industrialized countries with increased advantages by enabling them to actually continue polluting while still legitimizing their claims to be doing something about climate change? What is the space for maneuver of poor countries in the face of increased vulnerability to the impacts of climate change and constrained opportunities to seek economic growth? In other words, "how are we to understand policies implemented by subaltern states with severe internal and external constraints? How are we to interpret the ideology and practice of state sovereignty in global conditions that undermine it?" (Coronil 1997, 384). These are some questions that weave throughout the following chapters.

This book is therefore about transnational governance: how it is constituted, how it operates, what it produces. Specifically, it interrogates the processes through which policies to address climate change are designed and authorized through negotiations between policymakers, scientists, and activists operating simultaneously at multiple scales and throughout various jurisdictional levels. These include the international arena where the UNFCCC parties gather to agree on policies, national capitals where such policies are reframed, and the forests of Amazonia and Mozambique where some of those policies are reinterpreted and implemented. The following chapters circulate thus between the Amazonian state of Acre, other cities of Brazil, Maputo, and Zambézia in Mozambique, as well as Washington, D.C.

REDD and the New Carbon Governance

In order to examine how transnational governance constitutes itself and operates, I chose a particular mechanism intended to preserve forests, and that relies on a particular valuation of carbon that transforms this gas into a commodity to be transacted in specific markets. This form of valuing carbon is intended to reduce GHG emissions—the culprits of climate change—and follows from the assumption that if people are forced to *pay* for their emissions, they will act as a rational *homo economicus* and seek less-polluting alternatives. Accordingly, the same logic would apply to a country's level—where national authorities will seek ways to reduce emissions—and to industries, which will similarly invest in alternative technologies that cause less emissions.

However, this process of carbon valuation is more layered than the mere attribution of a price. What complicates this apparently simple idea, at least in part, is the way in which carbon markets have been conceived. As I will explain in more detail, what is traded in these markets is not actually carbon (or any other GHG), but atmospheric (empty) space (Machaqueiro 2017). What countries and industries purchase in carbon markets is, in fact, an authorization to keep polluting, in a way that fills that atmospheric space. This authorization was either previously created through negotiations by the different countries within the UNFCCC or generated through projects that allegedly reduce emissions already existent in the atmosphere (therefore compensating for new emissions).

One of these mechanisms that supposedly sucks emissions out of the atmosphere is called Reduced Emissions from Deforestation and forest Degradation (REDD). Since trees absorb carbon from the atmosphere, they act as "carbon sinks." When one tree is cut down, it releases carbon into the atmosphere, thus increasing emissions. But if that tree is maintained alive, its carbon will be kept. The logic underwriting REDD is, thus, that avoiding deforestation that was supposed to happen is equivalent to reducing carbon emissions. After being introduced, the acronym REDD acquired a "+" becoming REDD+. This shift was meant to signal that, besides avoided deforestation, REDD also includes the enhancement of carbon stocks. For simplification purposes, I use the initial acronym even though I am always considering the broader definition symbolized by the "+" sign. I need to emphasize that REDD is *not* about planting new trees that can absorb the already existing emissions of carbon in the atmosphere; rather, REDD is about preventing deforestation and forest degradation—or, even more accurately, reducing the rate of *forecasted* deforestation. Importantly, the accounting of the compensation that it can provide is directly dependent on a hypothetical scenario of future deforestation.

In theory, REDD would work as follows: as the owner of a few dozen-hectares property, I declare that I intend to remove timber from the property at a rate of 30 percent a year. That amount of timber I plan to remove is equivalent to two hundred tons of carbon released into the atmosphere (each ton corresponding to one carbon offset). Under REDD, if I agreed to deforest at a rate of only 15 percent a year, I would be paid for one hundred carbon offsets. The company paying me for those offsets would then either use them to be able to pollute one hundred tons of GHG or sell them to another company in need of more atmospheric space to pollute. As it has been presented in the various UNFCCC meetings, REDD is, in sum, a mechanism under which industrialized countries financially compensate countries in the Global South in exchange for the preservation of their forests.

Unsurprisingly, REDD has been contentious since its very beginning, raising questions about its real effectiveness in fighting climate change (Lohmann 2008), criticized as generating authorizations for escalating emissions, and increasing concerns over the possibility that REDD would just add to the persistent rationale of exploiting the Global South to the benefit of rich countries (Hoefle 2013). Such concerns stem from the fact that corporations would rather pay for avoided deforestation in poor countries than invest in less-polluting (and far more expensive) replacement technologies (DeShazo, Pandey, and Smith 2016, 70). Indeed, the several markets created to put a price on carbon have consistently rendered "forest carbon" cheaper than any of the so-called clean technologies. REDD has also been criticized for the conflicts that it can create among forest communities (Jindal, Kerr, and

Carter 2012; Larson et al. 2013; Beymer-Farris and Bassett 2012; Hein 2019), for exacerbating tensions around land tenure, property, and carbon rights (Mahanty et al. 2012; Asiyanbi 2016), for increasing inequalities (Howson and Kindon 2015; Chomba et al. 2016), and for being a form of neocolonialism[3] based on the commoditization of nature (Castree 2003).[4] Although discursively introduced as an innovative policy instrument that changes the tools to address forest conservation (namely through its market approach), REDD has, however, maintained older practices and shortcomings (Lund et al. 2017) dooming it to certain failure. As much as REDD can enthuse policymakers and conservationists for its claimed success by following the rules of markets, within a market logic, cutting trees is still more profitable than preserving them for their carbon sink capacities (Fletcher et al. 2016).

While my analysis is built upon the work of all these scholars, I am less interested in REDD's flaws than in its persistence and capacity to continue garnering support among policymakers, scientists, and environmentalists (Asiyanbi and Lund 2020). The explanation to REDD's persistence in the face of such apparently discrediting criticism cannot be found only in the "epistemic circulation" in which REDD advocates and practitioners find themselves recursively "selling the success" of this mechanism (Büscher 2013). It cannot be found either in the commonsensical idea that protecting forests is something inherently good. Rather, as I will detail, it can be found in REDD's operative flexibility, in what it enables and produces—especially in the context of transnational governance. So, why is REDD important?

The global governance of the problem of climate change—expressed in mechanisms such as REDD—presents some of the features that, in my view, characterize the new modalities of transnational governance. Although transnational governance (of which environmental global governance is part) is not new, it has changed significantly since the 1990s, enabling and expanding radical forms of neoliberalization in the countries under its intervention, namely through a level of capillarity that has not been reached by any other form of transnational governance. This capillarity—detailed throughout chapters 4 and 5—is the main feature of the new modalities of transnational governance. The second one is the couching of this governance in the authority of science. Even if the scientific discourses deployed to justify certain climate change policies are utterly simplified (as in the case of REDD), such policies are still granted greater legitimacy, thus foreclosing other policy approaches.

I want to make clear that while I am critically analyzing the ways in which the problem of climate change has been defined and how subsequently the policies to tackle it have been designed, I am not challenging the science that supports the existence of a problem called climate change. While I argue that the policies to tackle climate change are the direct outcome of a specific way of *socially constructing* this problem (Hacking 1999), I am not saying that climate change is not real. Instead, my effort here is one of renewed empiricism that can open new ways of understanding the problem and, with it, of formulating new (and more effective) policies to solve it.

The invocation of science in the context of climate change policies is not unique in the arena of transnational governance. Another example of how science can be deployed for policy-legitimacy purposes is the case of the "scientific discourses" that were (and continue to be) mobilized to justify austerity measures across Europe (particularly Southern European countries) as the only solution to the 2008 financial crisis—even in the face of increased depression due to the implementation of such austerity. The TINA ("there is no alternative") argument founded on supposedly scientific studies, such as the one conducted by Reinhart and Rogoff (2010), justified the political interventions led by the International Monetary Fund, the European Council, and the European Central Bank, imposing an austerity regime that only aggravated the structural problems of these countries, further impoverishing them.[5] As with the opaqueness of the financial world, the complexity of climate change offers a particularly apt context to deploy science to legitimize certain political options, while foreclosing others.

Finally, the third feature of the new transnational governance (and that is perfectly embodied by the purported goals of REDD) is its deployment of common spaces and/or values—in this case, the planet or the climate—to justify its interventions. This feature applies, as well, in transnational interventions to build democracy and human rights—concepts whose value is hardly challenged even though their exact meaning and associated practices are never fully clarified (Englund 2006; Coles 2007). As a result, and because very few hardly question the value of the planet or the importance of saving it, transnational interventions in the forests of the Global South become more than legitimate; they become an imperative.

In sum, by taking REDD as an entry point, I demonstrate that the new modalities of transnational governance justify its global approach on ideas of a shared space or values (like the planet, the economy, democracy, or security), and second, legitimize their interventions through a scientific discourse frequently followed by neoliberal principles of market rationality. While these processes are constantly and dialogically negotiated and re-signified at multiple levels (international, national, and local), they also involve conflicting constituencies and contradictory goals. Ultimately, these modalities of transnational governance not only depoliticize the objects of their intervention (and the interventions themselves), but also successfully divert issues of responsibility and political accountability.

The Social Life of REDD

If the idea of protecting charismatic forests such as the Amazon appears as intrinsically positive, such conservation efforts become even more valuable in the face of discourses explaining that emissions from deforestation are responsible for around 20 percent of global emissions. The planet needs to be saved! However, there are two arguments to be made about the role of avoided deforestation in the context of climate change policies: one related to the forests at stake when we talk about REDD and another concerning the genealogy of REDD. In order to make these arguments, the following chapters will carry out the ethnography of REDD's social life (Appadurai 1986). By social life I mean the actors involved in the processes of policymaking, circulation and implementation (i.e., scientists, policymakers, advisors, NGO practitioners, environmentalists, political activists, entrepreneurs, etc.); the networks that connect these actors to specific sites and among themselves; the scientific practices and discourses that inform these actors' beliefs and practices; the social rituals they perform (like the meetings where REDD is discussed); their system of beliefs and myths and how these circulate among them and between places; and, most critically, how REDD itself circulates, is negotiated, and acquires different forms, values, and meanings across places and among different actors.

In attending to the different aspects of REDD's social life it is clear that the deforestation that it is supposed to prevent is located in the Global South. Forests from industrialized countries are never mentioned in the overall forest carbon accounting, independently of whether they are being planted for industrial purposes or are native forests under threat by some sort of development enterprise. For instance, in 2016, when the government of Poland decided to open the Bialowieza forest to logging activities, the EU imposed a fine, but parties at the UNFCCC did not consider including Poland in a REDD scheme.[6] Similarly, while the Canadian province of Alberta continues to explore its tar sands at the expense of its forests, nobody suggested including Canada in a REDD pilot experience.[7] The focus on forests of the Global South is sometimes justified by controversial claims that tropical forests have greater sink capacities than other types of forests (Baccini et al. 2012; van der Sleen et al. 2015). However, such claims ignore the fact that many forests in the Global South are not tropical, raising questions about the worth of conserving certain forests in detriment of others, and the criteria used to make such decisions. The claims about the increased sink capacity of tropical forests also conveniently ignore that the major drivers of deforestation of these forests are directly tied to international trade that serves the demands of industrialized countries. Indeed, throughout history, the wealth of industrialized countries was partly built upon the extraction of resources and the destruction of forests from colonized territories. Some REDD discourses about the importance of tropical forests reproduce in many ways a form of exoticism associated with this type of landscape—and that can be traced back to the beginning of colonization—followed by (equally colonial) assumptions about the incapacity, or unwillingness, of poor countries to manage their forests and curtail deforestation (Neumann 1998; von Hellermann 2013).

When the problem of climate change emerged in the public domain and began to be mainstreamed by the media, forests did not have the relevance that they hold today in discussions about the solutions to this environmental crisis. In fact, in 1982, during a hearing in the US Congress within the Committee on Science and Technology—at that time presided over by Al Gore—the scientist James Hansen "had been irritated . . . by all the ludicrous talk about the possibility of growing more trees to offset emissions. False hopes were worse than no hope at all: they undermined the prospect of developing real solutions," explained *The New York Times* reporter (Rich 2018).[8] In the same piece, Hansen is again quoted, in a more recent statement, acknowledging that "most of the carbon absorption could be handled by replanting forests and improving agricultural practices," but only "if emissions, by miracle, do rapidly decline" (Rich 2018). In the absence of such a miracle, planting trees or avoiding deforestation continue to be ludicrous suggestions to solve climate change—and yet, the role of forests in curtailing emissions has only increased within the UNFCCC policy discussions since 2005.

The ideas of planting trees or avoiding deforestation were always present in the climate negotiations. However, many parties rejected them consistently, not just for technical reasons—the most important of all being their ineffectiveness—but also for political ones: investing in this low-cost solution would divert efforts to more needed measures to reduce emissions. As mentioned, the political context of the UNFCCC meetings gradually changed, enabling the inclusion of forests in the policy toolkit to address the problem of climate change. These changes were related to a greater propensity by policymakers to adopt market-based solutions instead of taxing and regulation (Harvey 2005, 160), especially after the several compliance carbon markets began operating in 2005—a propensity that follows from the financial habitus of decision-makers (LiPuma 2017, 101). Simultaneously, the idea that the so-called developing countries had to equally commit to fighting climate change became increasingly accepted. As such, while in 1992 the industrialized world acknowledged its responsibility and accepted that poor countries were entitled to seek economic growth without the constraints of reducing emissions, after the beginning of the 2000s that scenario changed. The tone of the negotiations shifted, and these countries were increasingly pressured to equally shoulder in the effort to reduce emissions. REDD emerged as the means to involve poor countries in such an effort.

REDD Within the Policy World

Despite not exactly being a policy, REDD depends on policy, shares some of its features with policy, and is inscribed in policy, thus demanding an examination of policy itself. Anthropologists have long problematized instrumental notions of policy that define it as a technical solution to a problem (see Shore and Wright 1997), that conceive the process of policymaking and implementation as unilinear and unidirectional (see Shore 2011), or that see in policy a cohesive and finished entity (see Clarke et al. 2015). Such critical conceptualizations of policy go against technocratic notions that not only deny the fact that policies are cultural and political products, but also erase the processes through which problems are defined as such in the first place (Mosse 2005). I take policy as a process in which discourses, practices, and relationships—involving multiple agents and institutions—frame a reality (be it problematic or not), and then define ways of intervening in it. This process is necessarily multiform and unfinished, being always subject to revisions, inflections, and translations (Clarke et al. 2015). By accepting the premise that policies, far from being objective or neutral, are ideological, political, and cultural products (Shore and Wright 1997), I examine the ways in which REDD is entangled in policy through three interconnected structuring elements: legitimacy, narrative, and value.

The relationship between these three elements operates as follows: the implementation of policies enabling REDD depends on the construction of a narrative that provides legitimacy (to both policies and REDD) and therefore attributes value to all of them. In order to provide legitimacy, the narrative has to not only stabilize the assumptions for policymaking (Roe 1994), but also promote the enrollment of as many parties as possible, which is achieved, in part, by orienting audiences to specific values that speak at different levels—local, national, and international—and that are overall accepted without being questioned. For instance, the value of planting a tree, promoting development, fighting poverty, or saving the planet are considered positive things in themselves; it is almost impossible to argue against development, welfare, or a healthy planet. Therefore, besides being endowed with these undeniable values that justify the decision to adopt a policy, the narrative has other performative features that promote enrollment and, therefore, produce legitimacy. The performance of this narrative is "founded on the widespread circulation via especially the public speeches of its high-status participants, leading to the collective and collectively ratified acceptance of the account" (LiPuma 2017, 219). In the end, the narrative crystallizes the conditions of future successes, or of future sources of legitimacy, by simply describing what constitutes success, or a source of legitimacy (219).

The narrative has depoliticizing effects (Ferguson 1994) not just in the sense of rendering the policies enabling REDD merely technical—as a technical solution to the problem of deforestation—but also in denying the politics of the object around which such policies are defined. That is, there is an *anti-politics machine* acting toward forests and the reasons why forests from the Global South are now at the center of climate change policies. Throughout my fieldwork in Brazil and Mozambique I lost count of how many times authorities and NGO practitioners from both locations deplored the fact that anti-REDD activists (the "No-REDDs") refused to discuss the mechanism in its technical terms, preferring instead to make it a political (or ideological) discussion. In fact, this depoliticizing effect is a crucial characteristic of transnational governance writ large: all international interventions in the Global South—be them conducted by state development agencies, NGOs, or transnational organizations such as the World Bank—are couched in technical and/or scientific narratives purportedly constructed in ways that preclude or neutralize potential challenges precisely by accusing such challenges of being ideological or of sustaining a political agenda.

Importantly, the narrative also performs the work of assembling (Li 2007). It assembles the different actors involved in enabling and promoting REDD to their different goals and agendas, the authorized forms of knowledge (265) underwriting REDD in its multiple forms (from the scientific discourses about carbon, to the knowledge deployed by practitioners), and the different and at times conflicting definitions of what REDD can be (Hein 2019)—all in a coherent representation. The more REDD enables the coexistence (and assemblage) of different and contradictory goals, the more effective and successful it is, precisely because it promotes the enrollment of more parties. Finally, the narrative provides a cohesive representation connecting policy to its practice, and the idea of REDD to its multiple and contradictory modes of implementation (Mosse 2005, 203).

When a policy is formalized into written form, that does not mean that such a policy is stabilized into a final format. The narratives about the policy can change, its legitimacy and justification can be reformulated in the face of different circumstances, and the value attached to it can also change. In short, the interpretation and implementation of a law are always mediated by social practices and the interests of the agents involved, according to their own position (Bourdieu 1993) in what I called the "REDD network." Both in Acre and in Mozambique the processes of policymaking related to REDD—including the inscription of REDD into state or national policy—went through different phases in which these three elements (narrative, legitimacy, value) were in permanent negotiation (chapters 4 and 5).

Parties at the UNFCCC created a legal template for the implementation of REDD that entails the creation of several policies in each country. But even if authorities are willing to transpose into national law a policy template created elsewhere, and the UNFCCC provide enough legitimacy to such a template, it is still necessary to provide this transposition with some form of local legitimacy. This is when the narrative is produced in a way that orients local audiences to specific local values. In the ethnographic cases depicted here, such values were directly tied to issues of authorship, ownership, and participation in the policymaking process. By stating that REDD-related policies were being created in participatory processes, having in consideration local circumstances and problems, authorities from Acre and Mozambique sought to legitimize their decisions, providing them with greater authority.

Narratives are fundamental to establish that authority and legitimacy by providing policies with a socioeconomic and historical background, a rationale for its need, and a depiction of its functionality. Furthermore, these narratives seek to ensure the continuous enrollment of people in policies and in leveraging their support by speaking to local values and assembling potentially contradictory elements while stripping them of their political content. Narratives are crucial to "sell success" (Büscher 2013) and in the confrontation of opposing narratives—as in the case of Acre—those presenting successful stories tend to dominate alternative views by being more effective in filling the gaps between claims and evidence (Roe 1994; Svarstad and Benjaminsen 2017), or by more effectively managing failures and contradictions (Li 2007, 265). The narratives on REDD are, thus, necessarily dynamic, changing according to the circumstances, but are mostly focused on its success—whether present or aspirational.

The value of a policy is usually measured through its outcomes, direct and indirect benefits, or the changes that it was intended to create. In the case of the policies enabling REDD, at the international level, for instance, their value concerns the possibility of creating a global forest carbon market, but also the overall benefits of forest conservation, management, and plantation. At the national and local levels, though, the value of policies related to REDD might be tied to development or the possibility of increasing the flow of international aid money. Given all the contentiousness around REDD, it is no surprise that it becomes indexed to multiple (sometimes ambiguous) goals such as sustainable development and rural development—both of which not only speak to local values and needs in Acre and Mozambique but can also include many different activities taking place in those territories.

Finally, another dimension of the value attached to REDD lies in the several meanings attributed to this acronym, which expand its hermeneutic scope and defy the initial (and apparently simple) idea of generating carbon offsets by keeping trees standing. That is why REDD continues to be talked about in so many different ways, from Washington, D.C., to Brasília, from Rio Branco to Maputo. For some, it is a mechanism for forest governance, or for the integrated management of a territory; to others, it is a development policy, leveraging other public policies, or even "just business!" REDD can ultimately be nonexistent, or dead before arrival, as I heard from my interlocutors in the field.

REDD's Circulation

REDD cannot be fully understood disconnected from the policies that make its implementation possible. The national and local policies that are created to foster REDD are very important, as they highlight how this mechanism is interpreted, re-signified and deployed outside of its negotiation platform. This does not mean that REDD, unlike its local instantiations, is coherently defined within the UNFCCC. Rather, it is the vague language used in this platform that enables REDD's polysemy, and thus, the success and continuous support of this mechanism, despite all the critiques and anticipation of its failure (DeShazo, Pandey, and Smith 2016). Even if REDD does not work, or its failure is confirmed, the idea of storing carbon in forests is consolidated, and will continue to attract policymakers, international donors, market advocates, and some environmentalists and NGO practitioners. Much of this attractiveness follows from the policies that countries put in place to enable the implementation of REDD (in its multiple forms). As such, I contend that REDD is very productive not just for the implementation of new forms of transnational governance, but also for national and local authorities in the pursuit of their own goals.

REDD cannot be fully understood either without analyzing the interconnectivity of the different levels across which it travels: international, national, and local. It is not just that REDD is different when talked about in the World Bank corridors or among NGO practitioners in the D.C. area, in comparison to what is being done in places like Acre and Zambézia; REDD's constitution and meaning are constantly changing due to the dynamic circulation between these three levels. International, national, and local levels are intrinsically relational to the extent that they continuously co-constitute each other in the designs, reinterpretations, and implementations of what REDD can be. It is through the continuous and interdependent relationships between the local, national, and international levels that REDD enables the pursuit of so many different goals under a coherent representation. It is also through these interdependent relationships that REDD facilitates the expansion of transnational governance.

But what are the implications of choosing REDD as a mitigation mechanism for climate governance? How is REDD circulating from the UNFCCC negotiation table into countries like Brazil and Mozambique? Having in consideration the conflicting and contradictory negotiation process that led to the inclusion of REDD in the climate change policy toolkit, it becomes clear that this mechanism cannot be understood outside of its sociocultural contexts of discussion, translation, and implementation. How is REDD being taken upon by Brazilian and Mozambican government officials and environmental practitioners? What does REDD look like in Acre? And in Zambézia, a poor rural province in the center of Mozambique? Ethnography provides, thus, a particular apt tool to capture and understand REDD's different forms and, at times, conflicting definitions as they circulate and unfold in everyday practices.

South-South REDD?

My initial interest in REDD in these two locations was spurred by reading about a REDD experiment taking place in Mozambique. In a publication named *South-South REDD: A Brazil-Mozambique Initiative for Zero Deforestation with Pan-African Relevance*, the lead author, Isilda Nhantumbo, explains that this "South-South REDD" started with a memorandum of understanding between Mozambican authorities and a Brazilian foundation, with the technical support of the International Institute for Environment and Development (IIED)—a British research NGO—the Mozambican University Eduardo Mondlane, one local NGO, and the Finnish forest consulting company, Indufor. The initiative was also supported by the financial aid of the Norwegian government (Nhantumbo 2012). In sum, a truly transnational endeavor, based on an old relationship between two Portuguese-speaking countries: Brazil and Mozambique.

Brazil's presence in Africa, and particularly in Mozambique, is not exactly new. But it has become more consistent over the past fifteen years since Lula da Silva won the presidency for the first time in 2002. In the 1960s, Brazil's initial incursions in the African continent were framed by the ideology of *lusotropicalismo* (Freyre 1933) and therefore based on a sense of brotherhood uniting Brazil to Africa and Portugal.[9] Brazilian diplomats at that time championed a notion they termed "racial democracy" to assert their advantage in the relationship with African countries recently liberated from colonialism, or still under colonial power, like Angola and Mozambique. The idea of racial democracy, a direct descendant of *lusotropicalismo* ideology, asserted Brazilians' moral superiority and lack of racism, unlike their northern neighbors in the United States. For Brazilian diplomats, their country was a natural link between Africa and the West, "two realities to which Brazil was deeply attached for historical reasons" (Cicalo 2014, 18). Therefore, Brazilians claimed to understand not only the experience of colonialism but also the needs of the colonized.

Highly ambivalent in its relationship toward Portugal, this Brazilian diplomacy nonetheless "used relations with Africa to assert autonomy from the US and stake a claim as an emerging world power" (Dávila 2010, 4). However, precisely because of the problematic relations that Brazilian rulers maintained with the Portuguese dictatorial and colonial power, and due to a lack of consistency in this foreign policy, Brazil's initial incursion into Africa gradually lost its significance before being abruptly terminated during the Brazilian debt crisis of the 1980s (Dávila 2010, 245). More recently, Brazil "reencountered" Africa and, once again, classified it as a "natural priority of Brazilian diplomacy" (Amorim 2013, 143) to the point that Brazilians who arrive in Mozambique claim to *know* Africa. During a conversation with a person working in the field of environmental consultancy and familiar with the problems caused by the arrival of the Brazilian mining company Vale,[10] I was told a story about Brazilians' *knowledge* of Africa: Vale distributed a questionnaire to Mozambican villagers with a specific question about the existence of *quilombos* nearby.[11] Puzzled by the word, many answered as if quilombos were a soccer team. According to this person, this type of cognitive arrogance was the major cause of the problems Brazilian companies faced in Mozambique.[12] This assumption (of knowing Africa) is based on Brazil's *blackness*, an identity feature openly admitted and deployed as a diplomatic tool by former president Lula da Silva (Cicalo 2014). In fact, during his mandate, Brazil's diplomacy in Africa not only deployed *blackness* as Brazil's advantage in comparison to other countries cooperating with Africa, but also apologized for the role Brazil had played in the slave trade (Cicalo 2014, 23).[13]

The country's more recent connections with Mozambique date back to 1995, when Brazil maintained a military contingent in the country mandated by the UN to keep peace after the civil war (Amorim 2013, 155), but were strongly reinforced throughout Lula da Silva's mandate. During his presidency, Brazil expanded its diplomacy in Africa with cooperation in the health and agriculture sectors, especially through the state-owned research company Embrapa, and established multiple agreements with Brazilian private enterprises such as Vale, and the oil company Petrobras. In 2013, technical cooperation represented 34 percent of Brazilian international aid to Africa, especially in Portuguese-speaking countries (Cicalo 2014, 22).

In the case of Mozambique, the importance of such technical cooperation did not mean, however, that Brazilian aid came unconditionally. After all, Vale has high stakes in the country, and the agricultural research company Embrapa has been involved in a very contentious development project affecting three Mozambican provinces.[14] Thus, this "rediscovery" of Africa by Brazil was described to me in a humorous but telling way by a Mozambican working for a Swedish NGO that specialized in small rural development projects. In his eyes, Lula came to Africa with the private sector to further the economic interests of companies like Vale and Petrobras but, unfortunately, he did not bring with him the unions or social movements like the MST.[15] At the end of the day, he added, Brazil is just another imperialist country like the United States.[16]

On Expectations About the Field

It was precisely this tension—between the assumptions inherent to what a "South-South cooperation" entails, and interventions that can be equated with neoimperialism—that captured my attention when considering how REDD was being implemented in Mozambique, in part, through Brazilian cooperation. During my first trip to Brazil, I was interested in both comparatively examining how REDD was being implemented in the small Amazonian state of Acre, as well as in how the terms of this South-South cooperation were being developed. Given Brazilian authorities' strong opposition to market-based approaches within the UNFCCC negotiations, my expectation was that this South-South cooperation would focus on the public governance of forests, preparing Mozambique to do so independently of all market-based initiatives. That is, I expected a cooperation that would challenge conventional schemes of North-South aid.

I wanted to understand whether this South-South cooperation might involve an alternative to the models that "developed" countries were proposing for governing the forests of the Global South, and what the implications of this reinterpretation of REDD might mean for transnational governance. The fact that Brazil had been implementing REDD at the federal level, and through broader policies not always related to forests alone (e.g., cash transfers to reduce poverty), was at least suggestive that in their cooperation with Mozambique, the Brazilians might be advocating a similar approach. Consequently, I expected Brazilian cooperation to be mostly concerned with preparing Mozambique to implement comparable public policies that connected forests with other sectors, rather than merely opening the country to the "projectification" of its forests' management (Meinert and Whyte 2014). I was also curious about whether this cooperation might promote Brazilian public policy experiments (Abdenur and Neto 2013) such as the Amazon Fund. This fund, managed by the federal government and supplemented by grants from foreign countries (notably Norway and Germany) has been used to curtail deforestation in Brazil through the implementation of different public policies in multiple Brazilian states; funding activities as diverse as cash handouts to prevent poor populations from logging and satellite imagery used to monitor illegal deforestation in the Amazon. The Amazon Fund was described to me by federal authorities as being Brazil's model for implementing REDD—one that refused the implementation of projects targeting carbon markets, and that was nevertheless funded by international donors to the extent that Brazil was able to show positive results.

My initial conjectures aside, the reality I encountered on the ground was in fact far more complex than I had ever imagined. First, in both locations I found a vast disparity between REDD as conceived within the UNFCCC and the multiple local reinterpretations of it, all of which challenged any possible assumptions of coherence in its conceptualization. Instead, REDD provided multiple localized occasions for the satisfaction of different fantasies. As I will show, it is precisely REDD's pliability that makes it such an effective instrument of transnational governance by opening the scope of intervention to multiple constituencies and agendas—what I call capillarity.

I also found that what I had thought of as a South-South cooperation developed at the governmental level was, in fact, a more complex entanglement involving the World Bank, Mozambican officials, and various Brazilian

NGOs. Even though this South-South cooperation turned out to be different from what I had anticipated, there was still a network of actors, scientific discourses and practices, and social rituals connecting Acre and Mozambique in unexpected ways. Ultimately, the ethnographic comparison of the ongoing processes of reinterpretation (and valuation) of REDD proved to be a more fruitful method of analysis, notably for what these reinterpretations can say about transnational governance writ large (van der Veer 2016)—transnational governance involves complex and contradictory processes of policymaking, interpretation, and implementation operating at and across multiple levels (international, national, local), in relation to, and interdependently from each other.[17]

Transnational Governance as Deep Connectivity

My analysis thus moves away from binary conceptions of center versus periphery (Wallerstein 2004), arguing instead that the REDD template created within the UNFCCC travels across a dense network connecting multiple actors, "sites, channels, arenas, and nodes of policy development, evolution, and reproduction" (Peck and Theodore 2015, 223). Instead of a hierarchical and unidirectional relationship between centers of power and poor countries, I see a complex network of actors, institutions, and practices intersecting three levels (local, national, and international) in which the effects of each actor's position-taking reverberate within and across all levels (Bourdieu 1993). This does not mean that there are not asymmetries of power throughout this network that make some actions and actors reverberate with greater effect than others—albeit not always in the way they might intend. Although the REDD template seeks some uniformity,[18] the policies enabling REDD's implementation in one place are not simple copies or transpositions of policies created elsewhere. Nor are the policies enabling REDD in multiple locations mere products of impositions by centers of power, be these donor countries or the World Bank. Throughout my research, I found that local and national actors continuously strive to expand their space for maneuver, manipulating and reinterpreting the REDD template in order to create opportunities for pursuing their own goals. Thus, although there are a lot of similarities between the REDD programs of Acre and Zambézia—due to the uniformity effect of the template—there are also specificities generated by the local reinterpretations and appropriations of the meaning of REDD, and of the policies through which it is supposed to be implemented.

As the REDD template travels across multiple places, its meaning is continuously remade through interactions between different actors, and its possible modes of implementation are renegotiated according to local circumstances and the goals of local actors, providing new valuations to what REDD can be and what can be achieved through REDD—paradoxically, by sometimes even denying it. The social trajectory of REDD experiments and policy models traced throughout the course of this ethnography are always mediated by brokers who may themselves move and act in multiple levels and across multiple social domains within particular levels.

Arnaldo, the World Bank team leader in Mozambique and a Brazilian national, is well aware of this reality, referring to himself as a translator between the world of the bank—focused on carbon trading—and the Mozambican world, a place he casts as "in dire need of development." Others, such as Roberto and Danilo, who are highly involved in the creation and implementation of the REDD program in Acre, also act as mediators themselves when they travel all the way from Acre to Washington, D.C., to showcase Acre's jurisdictional program to the "carbon folks" at the World Bank. In sum, REDD's circulation, translation, and resignification are not a mere dialogical process between center and periphery (see Escobar 1995), but a deeply multilocal and interconnected one in which actors, ideas, and practices intersect and co-constitute each other.

Although the history of REDD within the UNFCCC negotiations shows how countries from the Global South were almost forced into a position in which they too had to assume mitigation responsibilities by opening their forests to transnational forms of governance, that does not mean that authorities from these countries are merely passive implementers of policies dictated from on high, but rather have proven to be effective agents in their own right in this process as well. Authorities from Acre, Mozambique, and other countries and jurisdictions have also used REDD and related scientific discourses on carbon to pursue their own agendas. Ultimately, through its circulation, REDD continuously changes its definition, scope, and goals, depending on the actors invoking it, the places where it is implemented, or the objectives intended (whether stated or not) to be achieved through it. Therefore, if REDD has been conceived as a mechanism to compensate for the nonaction of deforesting, it can also be invoked to cut trees in what is considered a sustainable way, as in Acre; or to plant new trees for industrial purposes, as in Zambézia.

These apparently contradictory interpretations and deployments of REDD demonstrate the broad hermeneutic scope and operative flexibility of this mechanism, hence its persistence, and even success. In sum, REDD's success derives from its ability to simultaneously sustain contradictory imperatives while opening new levels of transnational intervention in the Global South.

Following the Network

In following this vast network generated around climate change policymaking, I had to make choices not just about the specific sites where I would carry out this multi-sited ethnography (Marcus 1995), but also about the kind of fieldwork and research methods I could employ—I had to cut the network (Strathern 1996). The opportunity to explore the rich political landscape of Washington, D.C., was provided to me by the location of my doctoral program. Indeed, the city hosts the headquarters of some of the most important environmental NGOs, not to mention the World Bank. I thus began conducting interviews in these organizations and attending conferences, talks, and other events very early in the program. I also took advantage of a network of contacts within environmental organizations that I had established in my previous job as a political analyst for the Portuguese government, to expand my access into the world of climate change governance.

In Mozambique, since REDD's implementation was being carried out with Brazilian cooperation, within a larger network of international collaborators, I felt the need to include both countries in my research. This was all the more so given the crucial role Brazilian authorities have had in the UNFCCC negotiations. Brazil is thus one of the sites of this ethnography, but in a multi-sited way, as well. On the one hand, I investigated the modes in which Brazilian national authorities have negotiated their position within the UNFCCC (frequently against the interests of countries like the United States, Japan, Norway, or the European Union); on the other hand, I followed the implementation of REDD in the state of Acre—itself contradictory to Brazil's federal position. This means that, sometimes, I compare the cases of Acre with Brazil, and of Acre with Mozambique (even though these two locations are not jurisdictional equivalents), and of Brazil and Mozambique in their national standing within the UNFCCC. In Acre, state authorities have great autonomy vis-à-vis the federal government, to the extent that some of their decisions were, at times, contradictory to the positions assumed by Brazil's government within the UNFCCC. In the case of Mozambique, there is a specific province (Zambézia) where the mechanism is being implemented, but decisions are made centrally, in the city of Maputo, and therefore, I more often refer to the country as a whole.

In seeking the best methods to apprehend transnational governance through the dynamic interconnections between these different levels of governance—international, national, and local—I adopted what Gusterson calls a "polymorphous engagement" approach, which means "interacting with informants across a number of dispersed sites, not just in local communities, and sometimes in virtual form; . . . collecting data eclectically from a disparate array of sources in many different ways" (Gusterson 1997, 116). As much as participant observation is the method par excellence of anthropology, performing it in the context of studying transnational governance and policymaking can be problematic, not just because of issues of accessibility but, more importantly, because it is hard—and occasionally impossible—to pinpoint exactly what and where to observe. Transnational governance and policymaking are everywhere and nowhere; they are as much visible as they are invisible; they happen both in the mundane practices of individuals and organizations and in the extraordinary events of political negotiations, as in the annual UNFCCC Conference of the Parties.

As such, my fieldwork also included a great deal of desk-based research, reading policy documents, briefs and reports, blogs, and news reports. I also spent several months in Acre and in Mozambique visiting government and NGO offices, either conducting interviews or just chatting with people; attending events related to climate change and the environment; and getting a better understanding of the personal stories of the people I got to know and their work over coffee or meals.

In Acre, although I spent most of the time in Rio Branco, I made two research trips to Xapuri—a city that was once the center of the state's development and known for being the place of birth of Chico Mendes (the political activist turned environmentalist) and of the former minister of the environment and now an international environmentalist, Marina Silva. I also had the opportunity to travel with a group of Americans auditing a project of forest conservation, which provided me with a privileged insight into how practices defined in an office of an environmental organization in the United States acquire a new life in a distant forest of Amazonia; and then, how that life is translated back into documents produced by auditors to be consulted once again in Washington offices. It also gave me a disturbing picture of how stereotypes related to the behavior of "gringos" in South America can be so accurate, while assumptions of a white superiority are surprisingly alive in the least expected places. This group of Americans kept questioning the skills and knowledge of the people they hired to take us into the forest by river, despite the fact that these guides have been living in that forest and navigating those rivers since they were born. And then, although part of the auditors' mission was to talk to the people living in the forest, for these American environmentalists, their prior (mis)understanding of the forests and of local conditions—shaped by their standardized approaches and quantitative methodologies—was clearly more valuable (and superior) than anything else they could learn from the people in the forest.

In Mozambique I spent most of the time in the city of Maputo, in large part because the country was suffering from a low-intensity war during my fieldwork, which precluded many road trips for security reasons. However, I had the opportunity to travel by plane to Zambézia, with a government delegation that included World Bank officers and members of a Brazilian NGO. This trip was important not just because it allowed me to see these people in action—explaining policies, negotiating them, convincing others of their value—but also because I got to observe them together, interacting with each other. More importantly, I had the opportunity to hang out with them, listening to their jokes and their personal complaints, including those against the corruption of the government of which they were part.

Finally, my fieldwork also included many emails, phone and video calls with people in many different locations, and archival research, both in Mozambique and in Lisbon. All of these different conversations emphasized the dynamic interconnectivities of the local to the national and the international, and vice versa, in this vast and fragmented network of climate change governance. It is precisely through the same sorts of interactions that transnational governance frequently happens.

Map to the Chapters

I start by highlighting the dialogical relationship between science and policy in the definition of the problem of climate change and, consequently, of its concomitant responses through the commoditization of carbon. I argue that policymakers operated a synecdochical re-inscription of climate science both by simplifying and partializing the problem of climate change. That synecdoche enabled the constitution of carbon as a commodity, allowing its trading, and authorizing the idea that it is possible to fix the whole by addressing one part.

In chapter 2, I provide an analysis of the UNFCCC negotiation process, showing how the language of these negotiations is essential to ensure the enrollment of parties in continuing to talk. As part of that language, I analyze the role of recursivity, reading obstacles, and the tropes of *transparency*, *accuracy*, and *consistency*, all of which establish the rules of the negotiation performance. The enrollment of all parties, with the subsequent agreement to keep talking, is one of the pillars of transnational governance, as it provides transnational organizations with the legitimacy to further intervene in poor countries, and to define the appropriate means of governance.

Chapter 3 focuses specifically on REDD within the UNFCCC negotiations, from its origins as a mere proposal to its formal inscription as part of the climate change policy toolkit. Focusing on issues such as the definition of forest or the equation used to measure the carbon content of trees, I show that despite the permanence of the same scientific uncertainties that prevented deforestation from being included in the Kyoto Protocol, the political context of the early 2000s propitiated the acceptance of REDD. The mechanism was pushed as a means to involve poor countries in mitigation efforts and as a form of expanding market-based interventions. I thus explore all the contradictions inherent to the setting up of markets as substitutes of policy regulation.

Chapters 4 and 5 turn to how REDD looks in Acre and Mozambique by ethnographically examining the policies that have been created in these two places, how those policies have been crafted and introduced, and then, how they have been contested and reformulated. The analysis of these two chapters is therefore anchored on the examination of policymaking processes around REDD and the connections between the local, national, and international levels of such processes. I argue that REDD is tremendously efficacious in maintaining a deceptively simple narrative about climate change that enables new forms of transnational governance. In order to understand these more recent modalities of transnational governance, it is necessary to examine ethnographically the different levels of the social life of a policy:

the international, the national, and the local. More importantly, those levels cannot be analyzed as separate or mere co-elements of a single policy process, but rather as a deeply dialogical compound, constantly in the making.

The conclusion summarizes the main arguments of the book, focusing on the distinct features of the new modalities of transnational governance enabled by the global governance of the climate. Throughout the chapters, I have sought to demonstrate the powerful role of multi-sited ethnography in the examination of transnational processes. While the comparison between the cases of Acre, Brazil and Zambézia, Mozambique underwrite the structurally flawed assumptions inherent in REDD's conception (instead of providing examples of misconceived implementation), the examination of the social life of REDD—across various jurisdictional levels and encompassing the multiple interpretations of what REDD can be—is what truly demonstrates the clout of this mechanism as an instrument of transnational governance.

Note on Sources

Throughout my research I have kept the identity of all my interlocutors anonymized, except when I was specifically asked not to. Whenever it was possible, I identified the organizations my interlocutors represented but only to the extent that it did not compromise their identity.

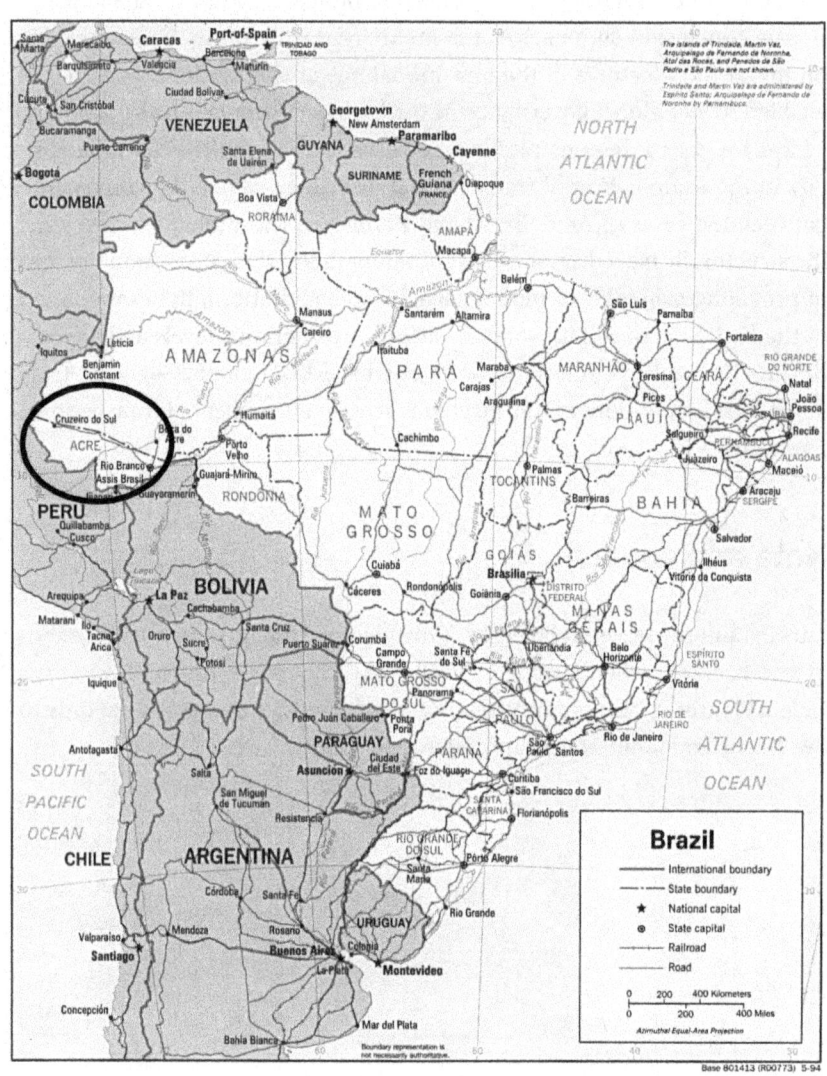

Map 1 Brazil / Acre. Courtesy of Central Intelligence Agency.

Map 2 Mozambique / Zambézia. Courtesy of United Nations Department of Field Support, Geospatial Information Section.

Chapter 1

Carbon and the Synecdochical Re-inscription of Science

> Indeed, the Earth's atmosphere is so thin that we have the capacity to dramatically alter the concentration of some of its basic molecular components. In particular, we have vastly increased the amount of carbon dioxide—the most important of the so-called greenhouse gases.
>
> The problem we now face is that this thin layer of atmosphere is being thickened by huge quantities of human-caused carbon dioxide and other greenhouse gases. And as it thickens, it traps a lot of the infrared radiation that would otherwise escape the atmosphere and continue out to the universe. As a result, the temperature of the Earth's atmosphere—and oceans—is getting dangerously warmer.
>
> That's what the climate crisis is all about.
>
> Gore 2006, 25–27

Across the spectrum of public discourse—whether in the media, policy discussions, or scientific fora—the problem of climate change has increasingly come to be narrowly associated with greenhouse gas emissions (GHG). These emissions are usually referred to simply as carbon, and the problem of climate change, in turn, as a process resulting from the excessive carbon emitted into the atmosphere because of human activity. Carbon emissions have come to stand for the entire problem of climate change—ultimately reducing its complexities—to the point that most policy debates revolve almost solely around questions about carbon emissions, such as how do we reduce them? By how much? At what costs? And by what means?

In this chapter, I examine the processes through which carbon has emerged as the dominant metric for measuring climate change, in both the science and the policymaking worlds, and how an entire regime of transnational intervention has emerged around what is ultimately a reductionist perspective of climate change—a regime focused exclusively on the simplified management of excessive GHG *as if this were the sole mechanism through which climate change should be addressed*. The definition of climate change in these terms has profound implications for transnational governance specifically by creating a form of global problem whose solution requires—and thus ultimately legitimates—particular forms of intervention. Such forms of intervention have been conditioned by both a co-production between science and international policy and, importantly, by its very early engagement with neoliberal governance.

This co-production implied the task of translating scientific knowledge to a broader and less technically proficient audience of decision-makers, which underwrote a tendency toward problem simplification, or what I call a synecdoche.[1] In other words, climate change and its causes have come to be dramatically oversimplified by many nonscientific yet highly authoritative actors. This simplification of climate change for nonscientists and their involvement in climate science has over time had a recursive effect on scientific practices and definitions themselves, informing their evolution in critical ways, namely, by providing opportunities for neoliberal principles to imperceptibly—and yet powerfully—shape the focus of climate science.

One crucial moment of recursive constitution of climate science was the development of the Intergovernmental Panel on Climate Change (IPCC), and its mandate to provide policymakers with comprehensive information on all issues related to climate change. The IPCC had a crucial role in rendering carbon emissions the primary object for transnational intervention, as well as in conceptualizing the climate as a singularity. Accordingly, the idea that actions undertaken in one part of the globe can offset those half a world away legitimized—with scientific authority—the creation of carbon markets as instruments to tackle climate change. The recursively reconstituted climate science that informs carbon markets has come to underwrite justifications for particular forms of intervention in the Global South by countries from the industrialized world. These new modalities of transnational governance are not only reproducing, but also extending already existing asymmetries between industrialized countries and the Global South—the hemisphere that will probably be hit hardest by climate change. While these new modalities of transnational governance are increasingly legitimized and expanded through the symbolic capital and authority of science, they are also instrumental in shaping that science itself, and in ensuring it remains responsive to neoliberal policy imperatives and values.

Human Intervention on a Global Climate

> Delta is helping customers travel a little cleaner today by offsetting carbon emissions on all domestic travel into and out of seven major airports with high corporate travel demand. . . . More than 170,000 corporate and leisure customers . . . will be accounted for through a Delta program that calculates the emissions per customer, then purposefully invests in global offset projects that provide social benefits while reducing emissions. . . .
>
> The carbon offsets purchased this month will be invested into three projects. . . . One of these is The International Small Group & Tree Planting Program (TIST), which empowers subsistence farmers in countries such as Kenya and Uganda to reverse the devastating effects of deforestation, drought, and famine through tree planting and sustainable agriculture. Meanwhile, the Kariba Project protects forests in Zimbabwe, while supporting the wellbeing of local communities.
>
> Delta News Hub, April 19, 2018[2]

The idea that the excessive carbon emissions in the United States can be somehow offset through their sequestration by forests in Kenya, Uganda, and Zimbabwe—half a planet away—presumes an understanding of the climate as a global phenomenon, a singularity where what matters is ultimately an aggregated effect. Thinking about the climate in these terms is a very recent notion. The understanding of the atmosphere as a single system, in which local events can have rippling repercussions elsewhere, is indebted to advances in multiple areas of knowledge developed over the last three centuries. Not only have advances in sciences like oceanography, geology, physics, meteorology, biology, and mathematics (among others) been crucial in enabling this understanding, but the concomitant development of powerful satellites and computers capable of processing increasingly growing sets of data have been equally important (Fleming 1998, 130). It was really after World War II (namely within the context of the Cold War arms race) that it became possible to articulate all these different forms of knowledge and technologies enabling the notion of a single, global, and interconnected climate, while also imagining ways of interfering in the climate at a planetary scale.[3]

These two ideas—that of a global atmosphere and the possibility of human intervention—became especially mainstreamed through the discovery of the effect of chlorofluorocarbon (CFC) gases in the ozone layer, and the growing awareness and concern with acid rains (caused by atmospheric pollution). Both vividly demonstrated the vulnerability of the planet to human actions and reinforced a more holistic vision of the planet (as an interconnected system of ecological relationships), while also putting in question longstanding assumptions about the supposed "automatic stability of biological systems" (Weart 2008, 106). These two phenomena also drove home the point that the atmosphere was not contained by borders and that human activities, decisions, and policies were consequential to environmental outcomes.[4] Both phenomena would, for the first time, also introduce another possibility into the equation: proactive human intervention. That is, should the planet not prove capable of either preventing or fixing damage to its ecosystems by itself, prevention and mitigation should—and could—be achieved by concerted human action and intentional design. The discovery of the ozone hole not only demonstrated in a very compelling way how fragile the atmosphere was to the inadvertent effects of human activity, but also placed the responsibility for, and possibility of, risk mitigation squarely within the realm of human agency. These new possibilities came together for the first time to show how "scientific findings about a future atmospheric risk could arouse the public enough to sway legislation" (Weart 2008, 123). No less than this is suggested by the National Academy of Sciences 1992 recommendation that "geoengineering"[5] should be developed to counter the effects of global warming (Kwa 2001, 135). The evolution of these two ideas—that of climate as a global phenomenon and of humans potentially intervening in it—have thus developed quite recently and in tandem with each other.

MAKING THE CLIMATE LEGIBLE: CLIMATE MODELING

These advances in scientific knowledge that occurred in the context of the Cold War's technoscientific race had one particularly important consequence for the evolution of core scientific notions and practices employed studying the global climate, namely the birth of climate modeling techniques. The practice of modeling the atmosphere through mathematical procedures had its origin within the Cold War technoscientific race and was maintained with research funding initially earmarked for militaristic purposes rather than environmental study per se (Demeritt 2001, 315).

Current knowledge about the climate is highly dependent on standardized data (in terms of format, space, and time)[6] and computer modeling. This means that the data collected by scientists (from the atmosphere and from the oceans) is parameterized and uniformized so it can be input into the modeling systems. These parameterizations and uniformizations aim at the standardization and quantification of complex data but can be, at times, controversial (Jasanoff and Wynne 1998, 9), especially when involving what is called "smoothing" the data (Edwards 2001, 46). This involves eliminating data that scientists consider anomalous, and the interpolation of intermediate values from known ones.[7] By removing less understood data and eliminating uncertainties, climate modeling has contributed to the growing idea that it is possible to know and understand the global climate and has consistently reinforced assumptions about the possibility of managing it.

Underlining the increased hegemony of climate modeling was the assumption that climate's behavior could be known—through data collected across the world by satellites and processed by supercomputers—and that, once known, climate could be predicted and regulated (if not controlled). The certainties and predictions of future scenarios supplied by climate modeling provide policymakers with the knowledge necessary for processes of decision-making and political action, even if scientists are sometimes aware of the uncertainties inherent in the process of modeling.

The overreliance of scientists on climate modeling has, however, been criticized both inside and outside the scientific community. This is because, first, these models and corresponding emissions budgets consider only the physical properties of GHG,[8] and thus ignore the socioeconomic context and meaning of them (Demeritt 2001, 316); that is, different GHG are measured equally without any attention to the activities generating them (be they industrial processes or cattle stocking, for example), nor to their different effects in terms of their radiative effects in the atmosphere.[9] Second, these models have been critiqued because they are also premised on the assumption that complex environmental systems can be decomposed into their constituent parts in order to infer the behavior of the entire system (Demeritt 2001, 317). This assumption is problematic in itself, for it takes the whole to be a mere sum of parts, thus neglecting the complexities arising from the combinations and relationships between those parts. Both these features of climate modeling could be equated to what Scott has called "de-skilling," that is, the "strip[ping] down [of] reality to the bare bones so that rules will in fact explain more of the situation and provide a better guide to behavior" (Scott 1998, 303) even though climate simulation often involves running models "without reference to any specific observed state" (Demeritt 2001, 317). Moreover, scientists have been perceiving these necessarily reductionist models as having a deterministic value; that is, the models are taken as instruments that actually predict what will happen instead of being considered as simulations that may or may not happen (Demeritt 2001, 318).[10] This is so because while scientists are well aware of the inherent uncertainty and contingency of their work, they are also "driven by institutional pressures toward expressing consensus in their findings" (O'Riordan et al. 1998, 370). Oversimplification of complexity and the dissolving of ambiguity is thus often a response to the pressure exerted by policymakers who "tend to exhibit a low tolerance for scientific uncertainty" (370).

In the process of generating and reifying certain specific ideas about the climate, modeling has also participated in the institutionalization of those ideas in ways that further invest them with authority. This has happened through the creation of an epistemic community that is not only structured around these ideas and concomitant forms of knowledge, but also that actively works to reproduce their legitimacy and thus, in the process, reinforce its own hegemony. Following Karin Knorr-Cetina (1999) and Caitlin Zaloom (2006),[11] I consider an epistemic community to be a group of people linked by shared forms of knowledge production, evaluation, and deployment into specific practices and informed by a shared set of values and norms—or a "habitus" (Bourdieu 1977). This community, which starts with scientists only, but later comes to also include other actors (such as policymakers and political activists) inhabits a social field—the *climate policy field*—which is structured by the distribution of capital and power among these actors, and in which each individual's position-taking is informed by the habitus produced by the social conditions of the field (Bourdieu 1993).

The forms of knowledge produced by this epistemic community are part of this capital and therefore, what imbues the community with power: first by transforming this knowledge into unquestionable axioms (e.g., the certainties provided by climate modeling) upon which practices are founded (e.g., institutionalization of carbon markets as the remedy to climate change)—all of which constituting a habitus (with a neoliberal disposition). Second, by deploying knowledge as a technique of their own reproduction as a community and informing individuals' position-taking inside the climate policy field (Bourdieu 1993, 42). In this case, this epistemic community not only reproduces these climate models and a simplified understanding of the problem of climate change (as well as the ideology and praxis involved in them), but also helps to shape the decisions and choices made in the field of climate policy. In this sense, the epistemic community is not only recursive—the knowledge it produces is determined by, and subordinate to, previous sets of knowledge also produced by the community—but also self-reflexive, as the conditions for its existence (and reproduction) are directly dependent on the knowledge it produces and deploys. This epistemic community thus exercises its cognitive authority in a way that strengthens its own ability to reproduce itself as a source of scientific authority.[12]

In a conversation with a climate scientist based in the Washington, D.C., area, I was told that scientists tend to present a consensual message about climate change so that their own uncertainties are not taken as evidence that climate change does not exist—a legitimate concern in the United States (Oreskes 2004). However, he did let me know that some of these less-known, highly complex elements of the climate include cloud radiation feedback, the melting of ice sheets and its effects on the sea level, and parts of the carbon cycle involving the oceans, which are not included in climate modeling. He also explained that the reasons for GHG's preeminence in the public domain (and policy world) have to do with the fact that this field of climate science is better funded, and that the issue of climate change is already so complex that including more factors could render it too complicated to be understood by policymakers and the public in general.[13]

Although the work of scientists can never be dissociated from their social conditions of production within the field of climate policy (Bourdieu 1993, 33)—for example, availability of research funds, solicitations from decision-makers, communication to audiences—there is a sort of co-misrecognition among this epistemic community. While climate scientists do not recognize themselves (and the knowledge produced by them) as part of a particular neoliberal political economy, policymakers also misrecognize the science upon which they are making their decisions. However, this politics of misrecognition is intrinsic to the field of climate policy, and to the authoritative capital of its epistemic community precisely because it impedes radical challenges to the knowledge it produces. In order to understand this politics of misrecognition, it is necessary to examine how climate change became the object of transnational political intervention.

Climate Synecdoche: From the Science Labs into the Policy World

The 1972 United Nations Conference on the Human Environment was the first international meeting bringing together governments and NGOs from different countries to discuss environmental issues, putting the environment on the international policy agenda. One of the most important outcomes of this conference was the creation of the United Nations Environment Program (UNEP)[14] in 1973, which would henceforth be involved in all global environmental issues. In the wake of the 1972 meeting, as research grew more focused in atmospheric modeling and climate simulations, the problem of climate change gained increased importance, inside and outside the scientific community, triggering a series of events that would determine the political approaches to this problem. In 1979, scientists from the World Meteorological Organization (WMO)[15] created the World Climate Programme "to coordinate and develop climate research and climate data" (Edwards 2001, 49). Together, the UNEP and the WMO organized several international meetings addressing the problem of excessive emissions and ultimately commissioned the *Brundtland Report* that would later (in 1987) alert governments and policymakers to the detrimental effects of increasing carbon dioxide emissions.[16]

In 1988, NASA scientist James Hansen—who first made headline news with the idea of global warming—testified before the US Senate Energy and Natural Resources Committee, asserting that global warming was already happening and that GHGs of anthropogenic origin were to blame (Allitt 2014, 245). Hansen's testimony and dramatic appeal to political action was important in calling public attention of the Global North to the problem of climate change. It also marked a moment in which climate change started circulating in the corridors of political officials. By the time of the 1992 Earth Summit, the problem of climate change was already starting to become mainstreamed among other scientists, policymakers, NGOs, and the general public, notably through the vernacularization of climate change promoted by the media (Callison 2014, 13).[17] With this mainstreaming, the initial epistemic community that was exclusively composed of scientists began to expand, including policymakers, diplomats, environmentalists, and activists from different NGOs.

One immediate consequence of the mainstreaming of climate change was the need to translate scientific information and make it accessible to policymakers. To that end, the UNEP and the WMO created the Intergovernmental Panel on Climate Change in 1988. The panel would study climate change in its various components through the examination of published peer-reviewed literature and produce assessment reports based on that information.[18] These assessment reports are intended to provide policymakers with "realistic response strategies for the management of the climate change issue" (Kutney 2014, 17) and supply policymakers with scientific information so they can make informed decisions. The creation of the IPCC and its hybrid character as a scientific-political body also responded to political concerns: if countries from the Global South were suspicious of reports prepared inside the Organization for Economic Cooperation and Development,[19] others—such as American conservatives—feared more radical suggestions such as banning the use of fossil fuels (Weart 2008, 152).

From the very outset therefore, the IPCC's official role as a forum for explaining science to policymakers was profoundly shaped by a less public transcript: that of ensuring that its prescriptions remain within political guardrails. In many ways, the work developed by the IPCC resembles that of government advisory committees whose members are frequently aware that "what they are doing is not 'science' in any ordinary sense, but a hybrid activity that combines elements of scientific evidence and reasoning with large doses of social and political judgement" (Jasanoff 1990, 229). The IPCC was structured in a way that enables it to accommodate different political concerns and imperatives through the authoritative language of science. Indeed, the IPCC and what it publishes is considered "the ultimate authority on climate change" (Callison 2014, 2); however, what is not clearly stated about IPCC's structure is that such scientific consensus is not merely a matter of scientific deliberation but is rather achieved through painstaking political negotiations. This negotiation process has led some authors to argue that the IPCC's scientific authority to speak on global policy issues is in fact based on a "rhetorical separation" between science and politics (Miller 2004, 60). Not only is that separation "rhetorical," but also scientists by themselves would never be able to leverage the needed authority to get political authorities of more than one hundred countries to sit down and discuss policies to address the problem of climate change.

More than researching the climate, or promoting advances in climate science, the IPCC's mandate is, in short, a direct response to political needs:

> The IPCC brings together the current state of knowledge on climate change science but placed in the context of *the practical needs of the policymaker*; critical issues are identified and the confidence of conclusions estimated. (Kutney 2014, 20)[20]

Although the IPCC has permanent members on its board, these are not necessarily scientists nor are they involved in the drafting of the reports. Invitations to experts to join the IPCC are usually communicated in advance to the respective governments, whereas "membership is restricted to nominations from a country to represent a national government" (Kutney 2014, 31).

Scientists responsible for the preparation of the assessment reports work on a voluntary basis after being recruited from several different disciplines and countries. The preparation of these reports involves a complex structure of authors including working groups' chairs and vice-chairs; coordinating, lead, and contributing authors; review editors; and government reviewers.[21] For example, expert lead authors (responsible for sections or chapters of assessment reports) are nominated by governments, individual scientists, and international scientific and industrial organizations and are ultimately chosen by the IPCC Bureau, while contributing authors (specialists in specific areas) are chosen by the lead authors. Since nominations are done by those already engaged with the ways in which climate change has been defined as an environmental and policy problem, those who do not adhere to such definitions are unlikely to be selected for nomination (O'Riordan et al. 1998, 369).

IPCC scientists' main task is to prepare assessment reports,[22] and with that purpose, "more than 1,000 of the world's leading scholars review more than 10,000 references every six years or so" (Kutney 2014, 19). The final result, "negotiated word by word," is the sum of scientists' findings unanimously endorsed by official government delegates from member states: it is not "mainstream science so much as lowest-common-denominator science" (Weart 2008, 156). Concretely, after the completion of the draft of an assessment report, it is further reviewed by government focal points, who are officials "assigned by the government . . . often bureaucrats in government departments and ministries" (Kutney 2014, 22). These bureaucrats make sure that the final text is acceptable by their respective governments, and areas where negotiations fail to reach a consensus are simply removed from the final report.

As the institutional hub of the climate change science epistemic community, the IPCC is the locus of the synecdochical re-inscription of science. This synecdoche occurs at two interdependent levels: at the level of scientists themselves, who choose what science gets to be assessed by them—a choice that can leave aside scientific perspectives on which it is deemed harder to reach a consensus; and which authors they want to work with. Subsequently, this process occurs yet again, at the level of bureaucrats and policymakers, who not only decide which people get to be involved in IPCC's work, but also, ultimately, on what the final report looks like—literally, word by word. These reports constitute a form of climate science's synecdoche because they oversimplify climate science and data (already simplified through climate modeling, itself based on data smoothing and parameterization) in order to render it digestible for policymakers. Such simplification is also a partialization that focuses on GHG emissions (further simplified into "carbon") while neglecting other factors—harder to translate into modeling language, such as cloud radiation feedback or the melting of ice sheets—that may equally influence the climate. Such partialization (or de-skilling), essential for the representation (and management) of climate, is also what conceals the ecosystemic complexity of climate, thus tapering potential solutions to the problem of climate change.

The IPCC's assessment reports are forms of re-inscription in the sense that they render the science visible (Latour 1987, 68), validating the new knowledge (Jasanoff 2004, 40), and making it reconsultable (Geertz 1973, 19) not only to scientists, but to policymakers and all transnational actors that will henceforth be part of the climate policy field. The reports are also a mechanism for expanding the reach of the epistemic community and their "enrollment" in this climate science synecdoche (Mosse 2005).

The IPCC epitomizes an ideal of policymaking that purports to be scientifically informed and therefore vested with greater authority. In doing so, the IPCC is also reinforcing current scientific conceptions of the climate as "a natural object to be understood, investigated, *and managed* on planetary scales" (Miller and Edwards 2001, 7).[23] That is, it is reproducing the type of climate science that constructs the problem of climate change as actionable by human beings through some sort of policy fix. The peer-reviewed science used by IPCC to prepare their assessment reports objectifies the climate in ways that enable its management, notably by rendering the climate measurable and legible. By simplifying an incredibly complex process that is still relatively poorly understood, and by reducing the problem of climate change to a question of GHG levels and circulation (itself cast in oversimplified terms with respect to both contributing sources and de facto effects), scientists have transformed climate into an object that is composed only of the elements they can isolate, understand, and measure (carbon and other GHGs) in ways that are amenable for input into their computer models. In this sense, they have transformed the problem into something that fits the tools they have—or have come to prefer—to use. In the process of feeding simplified, schematic, and standardized data into the increasingly complex computer modeling processes they use, however, scientists also set aside sets of relations and other complex elements of the climate that are not as readily codified, or that do not lend themselves well to the forms of understanding their value.[24]

The hybrid composition of the IPCC provides a fertile ground for the black-boxing of climate science, but also for the politics of misrecognition within the field of climate policy. By presenting the problem of climate change in isolation from other environmental problems and simplifying it into the axiom of excessive carbon emissions, the IPCC promoted the idea that managing climate change can be focused solely on the control of GHG emissions. The simplest element of scientific inquiry through climate modeling—greenhouse gases—thus became the target of policy intervention. Understood as anthropogenic, cumulative, and manageable, they have also become the unit through which interventions should be operationalized. Accordingly, the IPCC provides the guidelines on the appropriate methods to estimate GHG emissions, future emissions scenarios, and corresponding consequences for the planet. These guidelines, developed for the estimation of countries' emissions inventories, constitute an effort toward the standardization and simplification of methodologies for accounting and reporting GHGs—a necessary condition to manage the climate.

Once the relationship between the premises that climate change is a global problem and that it requires a transnational intervention targeting the reduction of GHG emissions was established, political leaders set in motion the negotiations leading to the decisions upon the modes and scope of such an intervention. The problem is that the very simplified way in which climate change has been framed, and the concomitant tapering of the science that investigates it, necessarily precludes the adequate and holistic responses that such a complex and severe problem require. It is to that political process of decision-making that I now turn.

Imagining a Market

The nearest thing that policymakers could find to a precedent for the problem of global climate change and a transnational intervention solution was the agreement that had been reached to address the CFCs gases that were causing a hole in the ozone layer. This problem brought a great number of nations together for a transnational political agreement to reduce pollution and phase out CFC emissions.

The commonality between these past approaches and the problem of climate change is the assumption by policymakers that the gases targeted by these policies are externalities. In classic economic theory, an externality is something not accounted for when planning an action (Pigou cited in Lane 2015, 41). Accordingly, the problem of climate change can be considered the result of a market failure, the result of not accounting for the costs of emitting carbon; that is, emissions are external costs of production. For classic economists, "the best response to market failure is not to abandon markets, but instead to use markets to reduce pollution in the most efficient manner" (O'Sullivan and Sheffrin 2006, 195). Carbon markets are thus supposed to correct this market failure by internalizing GHG emissions through the establishment of the appropriate price of this externality.

When world leaders got together to discuss the common goal of reducing GHG emissions—leading to the Kyoto Protocol—the United States insisted on a market approach, putting the idea of an emissions trading mechanism on the negotiation table very early on and arguing that it would be more effective than regulation, keeping costs low. The alternative—preferred by European countries—was a "command and control" approach,[25] including a tax to discourage emissions, which was rapidly dismissed as being "overly technology-prescriptive and innovation-unfriendly" (Yamin 2005, 4). Given the legal and political difficulties of imposing a new tax across the European Union, the creation of an emissions cap and trading mechanism became a far more attainable goal. The European Union Emissions Trading Scheme would thus become the main mechanism to internalize European emission in the production costs.[26]

In 1997, when the Kyoto Protocol was signed (see chapter 2), it included the creation of three flexible mechanisms, commonly referred to as "carbon markets," to help industrialized countries achieve their emissions reduction goals. One of the mechanisms, called International Emissions Trading

(IET), allows countries with binding reduction targets to sell their assigned number of units if they do not need all of them or buy from other parties if they exceed their emission quotas. It is, therefore, a simple market between emissions credits assigned to parties by the Kyoto Protocol, taking countries' needs and development trajectories into consideration. The other two mechanisms, the Clean Development Mechanism (CDM) and the Joint Implementation (JI), are project-based offsetting mechanisms. This means that by implementing projects in so-called developing countries, these mechanisms generate offsets that can then be acquired by parties (usually so-called developed countries) who need to comply with their emissions reduction targets. These projects can be, for example, the replacement of a coal-powered utility by a wind energy facility, the capture of methane in waste composting, the distribution of improved cooking stoves, or a forest plantation—anything that, according to IPCC's rules, is deemed to reduce emissions.

The credits or offsets of these three flexible mechanisms are fungible in the sense that they can be traded interchangeably among parties for compliance purposes. These mechanisms started operating in 2005, at the same time as the European Emissions Trading Scheme (ETS). This market follows the same logic of IET, in that it is a cap-and-trade system. This means that European authorities establish an emissions cap and the number of allowances distributed among emitters; then parties use the market to trade among themselves their assigned emission allowances, while also resorting to the project-based mechanisms (CDM and JI) to comply with the defined cap.[27]

In parallel to the creation of these four mechanisms, also referred to as "compliance markets" (because their purpose was to comply with reduction targets defined by the Kyoto Protocol), other emissions trading mechanisms were created by different market agents, with the purpose of responding to growing business concerns of corporate social responsibility. Although not legally mandated to reduce their emissions, some companies choose to offset their emissions in some of these voluntary markets so they can present themselves to consumers as environmentally concerned.[28]

Despite cost-efficiency arguments by American, Canadian, Japanese, Australian, Norwegian, and New Zealand authorities,[29] the prospect of implementing emissions-reducing projects in poor countries was initially not well received by the governments of these countries. Indeed, while industrialized countries contended that these projects would also be an effective way of involving poor countries in the effort of mitigating climate change while contributing to their sustainable development,[30] authorities from the Global South perceived the proposal as a form of "carbon colonialism" (MacKenzie 2009a, 149). That is, instead of regulating the industries in their own territories, industrialized countries would use poor countries' ground to (cheaply) compensate for their pollution, while also determining which development paths they would be able to follow or technologies to adopt.

Nonetheless, US negotiators were able to convince the rest of the parties that a market would be the most effective mechanism to deal with the diversity of GHG (and their different global warming potentials) as well as the multiplicity of emission sources (both in terms of sector and geography) by simplifying all these variables into exchanges of a single exchange unit: carbon. Significantly, while the problem of climate change was oversimplified through the creation of carbon markets, these markets are rife with bureaucratic and complex rules and procedures that render the understanding of their mechanics extremely hard, and therefore virtually inaccessible to public scrutiny.

The victory of the market approach over regulation demonstrates how the synecdochical re-inscription of climate science was informed by a market ideology that was predominant within the epistemic community involved in climate policymaking. Markets became the main transnational policy instrument to manage the global climate, and carbon emissions its commodity. Yet, as in all markets, there are rules of exchange that need to be created and followed according to specific theories of value.

Rules of Exchange

FUNGIBILITY

When establishing the mechanics of carbon trading, the first complexity to be dealt with was the amount of different GHGs with different warming potentials, which raised the issue of commensurability.[31] Global warming potential measures how much energy the emission of one ton of a GHG will absorb over a given period of time—usually one hundred years—relative to the emission of one ton of carbon dioxide (CO_2).[32] In the process of instituting a trading mechanism, the IPCC addressed this problem by declaring carbon to be the most important GHG. This effectively ignores any differences in effect or sourcing among gases in order to create a single quantifiable (and ultimately tradeable) unit—a "ton of carbon dioxide equivalent," or tCO_2e. Through this unit, all GHGs are invested with one form of value—their carbon equivalence—and their different warming potential is flattened and made commensurate with carbon dioxide's potential. As noted by scholar Donald MacKenzie, the global warming potential has been challenged as the best metric to estimate the effects of different gases in the atmosphere, and scientists themselves recognize the significant uncertainties entailed in the estimates of each gas's warming potential (MacKenzie 2009b, 446).

Even in terms of this simplified metric, the IPCC has always declined to specify what a safe level of GHG concentration in the atmosphere is, stating instead that the "stabilization of CO_2 concentration at any level requires eventual reduction of global CO_2 emissions to a small fraction of the current emission level" (Yamin 2005, xxxvii). However, it is the IPCC that defines the different gases' global warming potential in computations that are not only black-boxed but also arbitrary;[33] that is, they are entirely dependent on IPCC's scientific authority (MacKenzie 2009b, 446). MacKenzie explains that this black-boxing is crucial for emissions trading since negotiating an exchange rate for each gas transaction would reduce tremendously the liquidity of such a market, eventually rendering it impossible (446). That is, it would be tantamount to trying to establish a global barter system in which each transaction would have to be negotiated ad hoc. The fungibility between all greenhouse gases under the unit tCO_2e turned carbon into the indexical referent of all emissions trading mechanisms, and these different mechanisms into "carbon markets."

Yet, despite carbon's indexicality, what is being traded in these different markets is not carbon per se, but authorizations to emit more GHGs based on future projections of emissions built by computer models.[34]

SPACE ABSTRACTIONS

So let's say you fly on a commercial carrier from Chicago to Amsterdam. You would use one of the many online carbon calculators to determine your share of the CO_2 emitted by the aircraft ... you might find that you were responsible for 3.5 tons of CO_2 emissions. You would then pay a carbon offset company for 3.5 tons of CO_2 offsets. The company (or nonprofit organization) would then invest your money in a project meant to reduce greenhouse gas emissions and you get credit for a share equal to the carbon you were responsible for on your flight. Because climate change is a non-localized problem, it doesn't matter where the emissions are reduced. Greenhouse gases spread evenly throughout the atmosphere and reducing them anywhere contributes to protecting the climate.

Worldwatch Institute, 2007[35]

The constitution of these markets would not have been possible had the idea of a planet with a global atmosphere not already been instilled in the actors involved, from scientists to policymakers.[36] Indeed, a carbon market depends on a spatial abstraction, meaning that the carbon reduced in one place is seen as the same carbon emitted in another (Bumpus 2011, 622). But such spatial abstraction is also a product of the cost-benefit analysis supporting the market rationale: it is cheaper (for countries in the Global North) to reduce emissions in poor countries than in their own industrialized backyard (McAfee 2017, 47). This cost-benefit analysis puts carbon in a similar position as other commodities whose profitability is predicated on the asymmetries between the Global South and the Global North, and on the reproduction of such asymmetries.[37] For that reason, some authors have echoed poor countries' allegations about emissions trading, accusing carbon markets of constituting a form of "carbon colonialism" (Hoefle 2013; Bachram 2004) and of reinforcing conditions of "structural poverty" (Bumpus and Liverman 2008, 133). One critique of the spatial abstraction inherent to emissions trading is that of obscuring the significance of place, and the different impacts that pollution can have in different spaces (Lohmann 2008, 362). Moreover, such spatial abstraction conceals the relationships between the sources of emissions tied to the production of goods and the places where such goods are consumed.

RECURSIVE TEMPORAL ABSTRACTIONS

Carbon trading is also dependent on a temporal abstraction constructed through scenarios of future emissions. That is, policymakers decide how many units they assign to industries based on an established cap and set industries' predicted future emissions (itself based on an assumed development path and a desired reduction goal). The expectation is that industries can adapt to a decreasing allocation of emissions licenses. These caps are dictated through a political process that may draw upon and invoke science but is far from being dictated by science alone. Such is most certainly the case with the national inventories constructed on the basis of IPCC's rules and methodologies. Climate modeling is a participant in this political process inasmuch as "climate model projections have been driven by highly simplistic business-as-usual scenarios of human population growth, resources consumption, and GHG emissions at highly aggregated geographic scales" (Demeritt 2001, 312). Unsurprisingly, the allocation of emissions allowances turned out to be excessive in the case of European industries, with the majority of them (mostly energy utilities and cement industries) receiving more emission credits than they need, which allows them to turn a profit by trading those allowances in the market.

> Thanks to overly optimistic forecasts of growth and fierce lobbying by heavy industry the EU Emissions Trading Scheme (ETS) has failed to incentivise cost effective reductions in emissions and instead enabled some companies to profit from the scheme.... The top ten Carbon Fat Cats[38] share between them 240 million surplus allowances, equivalent to the annual combined greenhouse gas emissions of Austria (87M), Denmark (64M), Portugal (78M) and Latvia (12M).... Since most power companies buy allowances to comply with the ETS and pass on the cost of compliance to EU power consumers ... EU citizens are unwittingly paying a subsidy to oversupplied industrial sectors, which are able to sell their surpluses without investing in emissions reductions themselves. (Elsworth et al. 2011, 5–6)

In the project-based markets, project proponents are required to present an emissions scenario called "business as usual" (in which future emissions are sure to be higher than present ones) and argue that their project will reduce part of those future emissions. Proponents also need to prove that their project was not already being considered and therefore needs support to become a reality. This temporal abstraction is problematic in several respects. First, in assuming that emissions will always increase, these projects forgo any possibility that industries might actually take voluntary steps to reduce their emissions. If nothing else, this establishes a rather perverse incentive for industries to comply with that premise by rendering it more profitable for an industry to forgo investments in emissions-reducing technology until an emissions-reduction project comes along and pays them to do so.

Even more problematic are the premises that inform the construction of emissions scenarios themselves. An emission credit, or offset, is based on the idea of a future reduction before it takes place. This implies a forecasted base from which to define that reduction. Unsurprisingly, future scenarios are highly vulnerable to baseline forecasts that involve significant exaggerations of future emission rates (Bumpus and Liverman 2008, 136).[39] In the end, the anticipated—and thus future—reductions achieved through the projects must be reintegrated into a present-day market so another industry can buy them in the here and now, and emit accordingly, adding to the compounding abstractions that underpin carbon markets.

IN SUM, A SEMIOTIC PROCESS

The naturalization of carbon as a commodity is thus a semiotic process based on this concatenation of temporal disjunctures and spatial abstractions that occlude any potential place-significance and the arbitrary fungibility of GHGs into the unit tCO_2e. As a commodity, carbon presents all the characteristics identified by Marx ([1867] 1990): it has use-value (since it is used to offset emissions), it requires labor (namely the semiotic labor of rendering all GHG fungible in one unit, the operationalization of the required time/space abstractions, as well as, in the case of projects, the labor of implementing reducing emissions projects), and it has an exchange value (enabling its trading among different agents). My argument here directly contradicts the idea of carbon as a "fictitious commodity," as asserted by Beymer-Farris and Bassett (2012) following Polanyi's work ([1944] 2001). The idea of carbon as a "fictitious commodity" oversimplifies this semiotic process, rendering invisible the synecdoche underlying carbon's valuation as a new commodity within an old market logic that is predicated on the inequalities between countries. It elides the ways in which the commoditization of carbon is responsive to the imperatives of a neoliberal form of governance that has transformed an operational unit for modeling purposes into a commercial unit and how, by doing so, it has simultaneously reinforced a restricted form of doing climate science. Finally, it obscures the politics of misrecognition within the field of climate policy, and the co-constitutive relationship between the members of its epistemic community and the knowledge it produces.

Unlike other commodities, though, the value of carbon is not defined by the labor time congealed in it, but by a complex embroilment of policy decisions, weather events, energy consumption, and agents' expectations. The commoditization of carbon is thus the product of this synecdochical re-inscription of climate science, while deeply shaped by a market ideology—itself rooted in scientific discourses of rationality (LiPuma 2017, 17). The re-inscription that undergirds carbon trading leverages the symbolic and authoritative capital of science, but it no longer responds to the canons of scientific production. Instead, it is responsive to the political demands of asserting markets as a solution to climate change. As such, it is not surprising to verify that the valuation of carbon (as a commodity) mismatches—if not outright contradicts—the stated policy goals underwriting the creation of carbon markets; that is, tackling climate change.

A Theory of Value for Carbon: Price Versus Social Cost

Classic perspectives on markets state that prices are formed through the confrontation of the opposing interests of buyers and sellers exchanging commodities—each trying to maximize their interests. This means that, except for situations in which there are regulations in place establishing minimum or maximum ceilings, it is the balance between supply and demand that determines prices. According to classic economic theory, then, the market equilibrium is achieved "when the quantity of a product demanded equals the quantity supplied" (O'Sullivan and Sheffrin 2006, 69). What classic perspectives ignore, and several social scientists have pointed out, is the role of culture in markets (Callon 1998; Miller 2002; Lee and LiPuma 2002) and how social and local mediations can neither be separated from markets' mechanics (LiPuma and Lee 2012; Ho 2009; MacKenzie 2009a; Sahlins 1992, 2005) nor from processes of price formation.

Similar to financial markets—marked by an apparent intangibility and a greater distance between sellers and buyers—carbon markets are frequently referred to by their own agents and participants as operating under an economic and scientific rationality, stripped from any kind of sociality (LiPuma and Lee 2012). These agents also claim that financial markets are guided by the transparency of numbers (Zaloom 2006, 143) and provided by an agency totally separated from (and at times working against) markets' agents (Ho 2009, 11; LiPuma and Lee 2012, 295). Despite these claims, the formation of carbon prices and the inherent contradictions between those prices and the goals that were supposed to be achieved through that pricing process provides a good example of how markets in general are profoundly conditioned by political imperatives.

Policymakers set up emissions trading mechanisms with the goal of establishing a carbon price that effectively internalizes emissions. Through this internalization—that is, incorporating the cost of emitting carbon into the processes of production—policymakers expected to promote a change in behaviors, leading to less emissions. Their decision reflects an ideologically grounded faith in the inherent rationality (i.e., mutually beneficial exchanges between companies and lower abatement costs) of markets. Such faith is informed by scientifically authorized procedures (namely, allocation of allowances and approval methods for emissions-reduction projects) as the most efficient instrument for placing an exchange value, that is, a price, on carbon.

To the extent that these markets should incorporate all relevant costs, they should also incorporate the environmental and social costs of emitting GHGs in industrial activities. Accordingly, the price of carbon ought to reflect such costs, much in the same way that other prices materialize corresponding costs.[40] In fact, such efficiencies are required if carbon prices are to ultimately encourage the development and implementation of nonpolluting technologies by rendering the act of emitting carbon too expensive. However, an examination of de facto carbon prices reveals a different rationality at work, one which departs from the claimed efficient internalization of emissions so touted by the champions of market-based solutions to climate change.

Kyoto's flexible mechanisms and the European market began their operations in 2005, registering a consistent increase in carbon prices during the following year but, by 2008, the prices plummeted, in some cases by as much as 90 percent.[41] This radical downward trend was explained by market agents as a result of the financial crisis that slowed industrial activities, thereby reducing the need to acquire emission credits. However, the main reason for falling prices, especially in the European market, was "arbitrary after all, the result of political processes of [excessive] allocation" (MacKenzie 2009a, 168), challenging previous claims of scientific rationality and cost effectiveness. After 2008, carbon prices never returned to their initial amounts, with the World Bank reporting in 2016 that 75 percent of the covered emissions are priced below US$10 per ton of CO_2e (World Bank, Ecofys, and Vivid

Economics 2016, 11). These prices can hardly represent the social and environmental cost of emitting GHGs, a cost that was estimated to be US$41/tCO$_2$e by the Office of Management and Budget from the Executive Office of the President of the United States in 2016.[42] The social cost of carbon is an estimate based on three variables: how the climate responds to shifting levels of GHG emissions, how much harm global warming causes, and the impact of that harm on future generations.[43] This means that such a cost is directly proportional to the following policy options, namely 1) accepting (or not) the scientific discourse that links global warming with the increase of GHGs in the atmosphere; 2) assuming (or not) that the harm caused by global warming is a matter of care; and 3) accepting that future generations should not be harmed (or can be harmed).

The significant divergence between actual prices and what ought to be the real cost of carbon (defined as an externality) raises questions about the stated purpose of using emissions trading to reduce the amount of GHG concentration in the atmosphere. Such questions are reinforced by other studies, namely one led by the High-Level Commission on Carbon Prices that estimates costs of "at least US$40–80/tCO$_2$ by 2020 and US$50–100/tCO$_2$ by 2030" as a necessary condition to achieve the emissions reductions targets defined by the Paris Accord (CPLC 2017, 3).[44]

Shortly after the US Office of Management and Budget (under President Obama) announced its estimate for the social cost of carbon at US$41/tCO$_2$, a market consultant wrote an article stating his concern about the potential future link between current carbon prices and the social cost of carbon that this announcement anticipated. According to him, this linkage could be problematic for litigation reasons because once the social cost of carbon was considered the right proxy for the economic damages being done by CO$_2$ emissions, it would become politically unviable to issue free emissions allowances to utilities or to have officially sanctioned carbon prices that were far below the official social cost of carbon value (Trexler 2016). Indeed, although policymakers and market participants have frequently expressed the need to keep carbon prices high enough to reduce emissions, these concerns were never about matching the social cost of carbon with market prices. Arguments to support policy interventions intended to produce scarcity and increase carbon prices (like the decision made by European authorities to postpone the auctioning of 900 million allowances) were about the need to save carbon markets and ensuring their maintenance in the climate change policy toolkit

(Machaqueiro 2017, 88). In other words, the production of scarcity within carbon markets has been less concerned with reducing emissions than with the survival of these markets. As such, despite the persistent mismatch between carbon prices and the social costs of carbon, and markets' lack of effectiveness in reducing GHG emissions (Gilbertson and Reyes 2009), policymakers from industrialized countries continue to consider them the best instrument to curtail emissions (O'Sullivan and Sheffrin 2006).

Moreover, while the rationale for setting up carbon markets followed assumptions of scientific rationality, these same policymakers recognize that they need to do a better job at regulating markets, thus contradicting the premises of efficiency and self-regulation that they once mobilized to justify markets' creation. This was apparent during a meeting held in Lisbon in 2017 to celebrate the International Carbon Action Partnership's (ICAP) tenth anniversary. ICAP is an organization that brings together policymakers from regional, national, and subnational levels to share lessons on their emissions trading systems and promote their future linkage. While calling for "urgent action to control climate change," ICAP's final statement about the meeting stressed that "emissions trading can help ensure that the goals of the Paris Agreement are met in a *cost-effective manner.*" It added that

> a well-designed emissions trading system (ETS) puts an *adequate price* on GHG emissions, guarantees emissions are reduced at the *lowest cost*, and offers governments certainty that mitigation targets will be met

and that,

> Well-designed ETSs will be needed if governments are to achieve increasingly ambitious, long-term decarbonization targets in line with the goals of the Paris Agreement. Recognizing this, signatories will work to ensure their systems are effective in reducing emissions and send an *appropriate, long-term carbon pricing signal.*[45]

Without ever mentioning carbon's current low prices, the three-page statement emphasized, instead, that emissions trading provides emissions reductions at "the lowest cost" while setting an "adequate" price on emissions. Yet, what "adequate" means or how much is "appropriate" was never clarified. Again, inasmuch as these hardly reflect subscription to the properties of efficiency that are supposedly the raison d'être for market-based solutions, we are nevertheless invited to ask what other forms of benefits are gained from championing them.

According to carbon market analysts, prices are mostly affected by "the pattern of fuel prices" and weather as it influences the demand for electricity (MacKenzie 2009a, 163).[46] However, prices are equally influenced by policies and direct governmental interventions. If policymakers provide signals over the possibility of increasing the economic sectors covered by emissions trading, or the chance of greatly reducing the number of allowances freely allocated, carbon prices will go up due to an increase of demand in the first case, and of scarcity in the second. Also, if the number of project-based allowances authorized to compensate for emissions is reduced, the prices of these offsets will decrease because there will be less demand for them.

Ultimately, as Callon (1998) and other scholars have shown, pricing does not emerge from a simple relation between supply and demand, but from "a much larger constellation of factors that are not usually quantifiable" (Miller 2002, 226). This constellation of factors is what LiPuma and Lee called the "invisible sociality" embodied in market agents, institutions, and the structure of financial practices (2012, 292), all of which are determinant in price formation and in the mismatch between carbon prices and the environmental and social cost of carbon. As such, the commoditization of carbon might have never been, after all, about the claimed need to internalize GHG emissions but rather an ideological response to the political needs posed by the problem of climate change.

Divergent Metrics of Value

The recent Paris Climate Accord does not establish an international carbon market to replace Kyoto's flexible mechanisms, nor does it define a carbon price. In fact, some noticed, the word "market" is never mentioned, something that is not accidental (Marcu 2016, 6). Given many countries' opposition to market mechanisms (a debate covered in subsequent chapters), the language of this agreement is purposefully vague. In its vagueness, however, it enables the creation of multiple subnational markets (with possible future linking in an international carbon market, if parties wish to do so), while maintaining the action of trading as merely voluntary. More specifically, the document establishes that countries can trade their international mitigation outcomes (ITMOs)[47] among themselves to deepen the reduction targets that they have set for themselves in their nationally determined contributions (NDCs), which opens the opportunity for a future market (Zwick 2016). As in the Kyoto negotiations, the Paris Accord exposes the logic underlying industrialized countries' positions. That is, based on a market ideology and cost-benefit analysis, authorities from these countries have consistently sought to reduce their emissions through commercial exchanges (not actions like technology shifts) and, more importantly, at low costs. If Kyoto's flexible mechanisms (and now the voluntary trading of ITMOs contained in the Paris Accord) meet industrialized countries' first goal, the inclusion of REDD and other forms of land management in actions to reduce emissions fulfills the second.

This situation demonstrates the sociality of markets—constantly denied by market agents (LiPuma 2017, 13–14)—and the inherent contradictions between the claimed objectives of policymaking and the intrinsic goals of market agents. On the one hand, the policy world responds to the predicament of governing the global climate attending simultaneously to two different constituencies: one that considers tax and regulation bad and markets good and that is not willing to pay the costs of their lifestyles, and another that demands action to mitigate climate change. On the other hand, market actors seek the lowest costs and the greatest profits, and the maintenance of the mechanisms that allow them to make those profits (i.e., markets). Despite the apparently divergent metrics of value—of policy and of the market—policymakers and market actors inhabit the same social field of climate policy, in which each actor's position-taking is defined in relation to other actors' position-taking (Bourdieu 1993, 30). Hence, policy goals of responding to their constituencies can indeed be compatible with the profitability goals of market agents.

Throughout the negotiations leading to Kyoto, the metrics of policy were often subsumed by the metrics of market, transforming the so-called flexible mechanisms into the primary instruments to reduce emissions instead of enforcing regulation toward the adoption of less polluting technologies. After Kyoto, this trend was accentuated through the rising number of regional market mechanisms, and the increased role of forests and the land sector as a critical part of the available instruments to curtail emissions. Afforestation, reforestation, REDD, and agricultural projects (among others within the land sector) are arguably cheaper than shifting technologies in industry, or developing renewable energy, besides offering profit opportunities from the crops involved and the opportunity to access land cheaply for other purposes. In this context, the price of carbon gets more entangled with other markets (food, cosmetics, timber, textile, financial, etc.), variables like access to land or capital, and political decisions not necessarily related to climate change or the environment. The goal of internalizing emissions in production costs (which justified the naturalization of carbon as a commodity in the first place), was thus overridden by the goal of providing reduction emissions choices at the lowest cost possible, to the point of a great mismatch between carbon prices and the social cost of carbon.

This seemingly paradoxical situation is best understood when carbon markets are examined outside of their "moral and ideological system" (Miller 2002, 224) according to which markets are the best instrument to give a price to carbon, and are instead seen as a means to achieve policymakers' unstated goals: to be able to claim that emissions are being reduced, and to do so at the lowest financial and political cost possible. The predominance of these policy goals can be seen as a direct consequence of the synecdochical re-inscription of climate science, through which carbon and its commoditization became hegemonic within the tools to address the problem of climate change.

These unstated goals of policymakers from industrialized countries explain why REDD attracted so much support among them (and market agents) when initially introduced. Despite the obstacles in rendering this mechanism an international compliance market, efforts toward that goal have not lost their vigor. A good example of the creativity involved in some of these efforts is the REDD+ Acceleration Fund, promoted by the Rockefeller Foundation and the Environmental Defense Fund (EDF),[48] and whose presentation I attended during an event in Washington, D.C. Every year, the NGO Forest Trends and Ecosystems Marketplace launches its report on forests and finance, identifying trends in projects and prices.[49] At the 2016 event, guests were hosted at the law firm Baker & McKenzie LLP, a sponsor of that NGO. During the presentations, the speakers agreed with the idea that the Paris Accord fortified the role of forests in carbon-reducing commitments, but they also insisted on the need to connect the public and private sectors in order to bridge the financial gap needed to implement REDD. The greatest concern of all these actors is the (so far) lack of success at including REDD within compliance mechanisms. But the vague language of the Paris Accord opened that possibility, some noted. A member of Baker & McKenzie thus introduced EDF and Rockefeller's idea to address that problem through the acceleration fund. According to him, it is necessary to look at the macro policy signals and bring the private sector in, as a precompliance mobilization of capital. The private sector, he stated, is willing to risk money if public finance can serve as a mitigation risk. More specifically,

> the proposed strategy is structured like a traditional investment fund. Corporate precompliance investors (expected to be companies which have carbon emission caps, such as airlines or utilities) invest in an option to purchase REDD+ credits in the future but only if compliance markets are accepting REDD+ credits at that time. (Edwards 2016)[50]

According to the member of Baker & McKenzie, in 2020 these credits would be recognized, and their value increased, spurring their trading. If, on the contrary, there is no market for these credits, then there would be a trigger mechanism to compensate all the investors.[51] At the end of the day, he added, there will always be a last-resort buyer: the states. In sum, the new acceleration fund intends to "encourage private investment into avoiding tropical deforestation by significantly reducing the risk of investing in REDD+" (Edwards 2016). It was never mentioned, however, why forest carbon credits are so important to keeping the costs of reducing emissions low or, from a different perspective, how forest carbon can enlarge some companies' profits in a new financial market, where risk is totally supported by public funding.

Similarly, a recent article published in the journal *Climate Policy* argued that since there are signs that "a large-scale international forest carbon market" is coming into place,

> enabling options-based transactions could minimize risks for both REDD+ buyers and sellers, help smooth the transition to future climate policies, and unlock important emissions reductions that can accelerate climate action in the context of significant policy uncertainties. (Golub et al. 2018, 3)

The authors justified the argument with the fact that

> compared to investments in abatement in fossil-energy and industrial sectors, this cost structure of avoiding deforestation is likely to be less front-loaded, with fewer and smaller initial investments and more recurring investments. (Golub et al. 2018, 5)

In other words, it will always be cheaper to avoid deforestation than to shift technologies. While many question the future of schemes like REDD (DeShazo, Pandey, and Smith 2016; Fletcher et al. 2016; Lang 2016; Lund et al. 2017), anticipating its gradual disappearance or transformation into something else (not necessarily market-based), forest carbon seems to be increasingly valued as the means to achieve low-cost emissions reductions, even if the reduction of emissions through REDD is no less than flawed.

Examining the expansion of schemes like REDD beyond the normative idea of internalizing carbon emissions by providing carbon with a price unveils a more complex picture in which scientific discourses are deployed in very specific ways to legitimize certain actions. The synecdochical re-inscription of climate science has not only enabled the erection of carbon markets as the hegemonic tool to fight climate change, but more importantly, it has produced a successful narrative about the potentialities and cost-efficiency of REDD (and other land-based markets) in reducing emissions. And as carbon markets expand into new domains (like forest carbon, or the financial sector), the metrics of value that once undergirded these markets' creation gradually shift. The claim that markets would be an efficient means of internalizing the costs of emissions has instead opened the potential for new forms of intervention in the territories of poor countries; that is, as long as there is a claim to be made about the possibility of reducing carbon emissions in the lands of countries like Mozambique or Brazil, there is a reason to manage those lands to the benefit of the global climate.

Conclusion: Recursive Synecdoches

The translation of climate change into the policy world has implied its utter simplification in a process I called synecdochical re-inscription of climate science. This synecdochical process was already part of the ways in which climate science constituted itself; notably, by its focus and reliance on climate modeling, itself built upon parameterization and data smoothing. However, such a process was greatly reinforced with climate change's expansion into the international political agenda and public mainstreaming: the more policymakers ask scientists for certainties about an extremely complex object—the planet's atmosphere—the more scientists rely on computer modeling, blurring the differences between predictions and simulations, between possibilities and probabilities; the more decision-makers demand knowledge, the more scientists' work is shaped by those demands, contributing to the co-misrecognition of this large epistemic community. Given that climate science was already partially built upon a simplification of its object of inquiry, then, the political synecdochical re-inscription of science is a recursion of previous synecdochical processes. The outcome of these recursive synecdoches is the partialization of this type of knowledge to the point of equating the entire problem of climate change solely with carbon emissions.

The presentation of the problem of climate change in this simplified way serves two purposes. On one hand, it follows the conventional logic of "development" interventions according to which the problems addressed by such interventions are considered in isolation and disconnected from broader structures of causality; problems are therefore represented in accordance with the tools at the disposal of the "development" industry (Ferguson 1994). As such, the problem of climate change is presented as liable for similar "development" interventions, which just have to be scaled up in size (because it is a global problem) but not in complexity (since that would require situating climate change within a broader network of interconnected environmental problems and within complex capitalist systems of production and consumption). On the other hand, this presentation enables the neoliberal instruments and logics that caused climate change to be involved in finding the solutions to it, therefore contributing to neoliberalism's reproduction (Fletcher and Büscher 2017)—such is the case of carbon markets. As Naomi Klein put it, "Climate change was presented as a narrow technical problem with no end of profitable solutions within the market system, many of which were available for sale at Walmart" (2014, 210).

Informed by a neoliberal ideology and mobilizing the authority of science, policymakers from the industrialized world transformed a scientific unit of measure for climate change into a unit of intervention. By declaring climate change a market failure, carbon emissions became the sole target of such intervention and carbon markets the best policy instrument to tackle them. Simultaneously, and because the atmosphere is global and single, these markets could operate emissions reductions in any part of the world to have a beneficial impact on the climate as a whole. The idea that it is possible to fix the whole (the planet's climate) by addressing one of its parts (say, a forest in an African country) is another outcome of this synecdoche.

Perhaps the most important outcome of the synecdochical re-inscription of climate science is how it shores up transnational governance in new ways. The idea of a global and actionable climate creates new forms of interdependency that legitimize different forms of intervention, because ultimately, what is at stake is the entire planet. If traditional forms of intervention were about elevating poor countries to the standards deemed appropriate by countries from the Global North (be it in terms of development aid, humanitarian assistance, or the spread of democracy and the rule of law) under this new logic, the goals to be achieved with these interventions are presented as concerning to the entire planet and its survival. As such, under the legitimizing imperative of saving the global climate, countries from the Global North accrue further authority to intervene in the forests of Acre or Zambézia, for example, while benefiting from the direct implications of such interventions. In other words, while claiming to be doing something about climate change, authorities from the so-called developed countries decline to enforce painful regulations that would de facto reduce the amount of emissions generated by their industries. In this sense, these kinds of interventions can be seen as a form of internalizing "development" in the domestic policy and economies of these developed countries in a cost-efficient way, since the resources spent in the reduction of emissions in the Global South are undoubtedly lower than the costs of intervening domestically.

Expanding mitigation initiatives to the forestry sector in the Global South has been the most effective form of ensuring these lower costs, hence the insistence of industrialized countries' officials in pursuing mechanisms such as REDD. Claiming to be protecting forests in developing countries and, thus, curbing emissions, authorities from rich countries can maintain their industrial activities and energy consumption patterns unchanged. Simultaneously, this focus on forests and the land sector in the Global South opens opportunities for the expansion of transnational governance into other domains of these countries—a process that I call capillarity, explored in chapters 4 and 5.

In sum, if this synecdochical re-inscription of climate science is an essential process to justify and legitimize new modalities of transnational governance (by providing it with the scientific legitimacy and urgency to operate and expand to the benefit of the planet), it is also responsible for an oversimplification of the problems to be governed, and therefore, for the curtailment of the range of policy possibilities to solve them.

Chapter 2
The UNFCCC Negotiation Process
Agreeing to Keep Talking

Introduction

Every year since 1992, authorities from more than 150 countries have convened under the United Nations Framework Convention on Climate Change (UNFCCC) to discuss the problem of climate change and debate possible solutions. These weeklong and bureaucratically complex meetings—called Conference of the Parties (COP)—involve dozens of national delegates, scientists, environmentalists, NGO practitioners, and officials from organizations such as the World Bank, and are a key element of what I call transnational governance. The global climate is addressed through the agreements reached in this transnational setting gathering hundreds of people. Indeed, each COP constitutes a complex ritualistic moment performed through language, during which both the format of the negotiation process and the subsequent documents agreed upon sustain and reinforce the legitimacy of the UNFCCC as the organization governing the climate.

The format of the negotiation process is the key to the continuous enrollment of all parties, who agree to keep talking, and in doing so, implicitly conform to the terms used to define the problem of climate change. Accordingly, the texts coming out of each COP, characterized by their vagueness, formalized language, and specialized jargon, frequently seem to decide very little. Yet they are crucial to the constitution of transnational governance through the formation, stabilization, and legitimation of a specific language through which such governance exercises its authority.

The Warsaw Framework for REDD+ was agreed upon during COP19, held in Warsaw in 2013, and is considered the legal foundation for the implementation of REDD. Although this document was hailed as a great achievement in the history of the UNFCCC's negotiations, it did not trigger the implementation of REDD on a large scale nor did it define concrete steps to establish a global market for forest carbon offsets as was expected by some of the parties and by various environmental NGOs. But it is precisely this paradox of having an agreement hailed a success that does not define concrete actions to be taken by the signatory parties that I will tackle in this chapter. To do so, I use the documents that emerged from COP19 as an entry point into the negotiation process itself and its performativity. By analyzing the role played by the vague language used in these texts, I offer a perspective on why such vagueness is crucially constitutive of the claimed negotiating success.[1]

Table 1 UNFCCC Conferences of the Parties

COP	Year	Place	Achievements
COP1	1995	Berlin	Agreement on the Berlin Mandate establishing a negotiation process to strengthen emissions-reduction commitments by developed countries.
COP2	1996	Geneva	Formally accepted the scientific findings on climate change proffered by the IPCC.
COP3	1997	Kyoto	Adoption of the Kyoto Protocol.
COP4	1998	Buenos Aires	Adoption of a two-year Plan of Action to devise mechanisms to implement Kyoto.
COP5	1999	Bonn	No major conclusions.
COP6	2000	The Hague	The US proposed to allow emission credits from forests and agricultural lands. No agreements. COP suspended.
COP6	2001	Bonn	Agreement on the operational rulebook for the Kyoto Protocol (flexible mechanisms, carbon sinks, compliance, and financing). President George W. Bush rejected the Kyoto Protocol.

Table 1 *continued*

COP7	2001	Marrakesh	Establishment of the Marrakesh Accords formalizing rules for the International Emissions Trading, the Clean Development Mechanism, and Joint Implementation.
COP8	2002	New Delhi	Adoption of the Delhi Ministerial Declaration calling developed countries to transfer technology to developing countries.
COP9	2003	Milan	Agreement on the usage of the Adaptation Fund to support developing countries.
COP10	2004	Buenos Aires	Adoption of the Buenos Aires Plan of Action to promote the adaptation of developing countries to the impacts of climate change.
COP11	2005	Montreal	Kyoto Protocol entered into force in the beginning of the year. For the first time the COP is held in conjunction with the first Conference of the Parties serving as the Meeting of the Parties (CMP1) gathering Kyoto's signatory parties.
COP12	2006	Nairobi	The Subsidiary Body for Scientific and Technological Advice (SBSTA) undertakes program to address impacts, vulnerability, and adaptation to climate change, later originating the Nairobi Work Programme.
COP13	2007	Bali	Adoption of the Bali Road Map launching a new negotiation process to address climate change, as the commitments agreed under Kyoto are due to expire in 2012.
COP14	2008	Poznan	Launching of the Adaptation Fund to support developing countries and the Poznan Strategic Programme on Technology Transfer.
COP15	2009	Copenhagen	Expected to reach a major agreement to replace Kyoto, turned out to be a disappointing document despite the presence of President Barack Obama and other important heads of states. Developed countries pledged up to US$30 billion in fast-start finance for the period of 2010–2012.

Table 1 *continued*

COP16	2010	Cancun	Establishment of the Green Climate Fund, the Technology Mechanism, and the Cancun Adaptation Framework.
COP17	2011	Durban	Parties agreed to reach a new climate agreement (to replace Kyoto) by 2015 for the period beyond 2020.
COP18	2012	Doha	Adoption of the Doha Amendment that establishes a second commitment period of the Kyoto Protocol. Parties agree to speedily work toward a new comprehensive agreement.
COP19	2013	Warsaw	Agreement on the legal framework for REDD, as well as a mechanism to address loss and damage caused by long-term climate change impacts.
COP20	2014	Lima	Before the Lima COP, Ban Ki Moon hosted a summit in New York, inviting governments, business, finance, civil society, and local leaders to mobilize action in advance for next year's COP in Paris.
COP21	2015	Paris	Agreement of a new climate accord to combat climate change.
COP22	2016	Marrakesh	Progress on writing the rulebook of the Paris Agreement.
COP23	2017	Bonn	Although taking place in Bonn, the meeting was presided by the president of Fiji.

Sources: unfccc.int/timeline; Wikipedia.

The UNFCCC and the Need to Govern the Climate

During the 1980s, the issue of "the environment" gained traction in the world of development, especially after the release of the *Brundtland Report*[2] in 1987, which compiled a list of ecological threats to the planet—including the dangers of CO_2 emissions—thus increasing political attention to the problem of climate change (Kutney 2014, 6). With climate change at the forefront of the transnational political agenda, in the 1990s the United Nations' secretary general decided to create an intergovernmental body to deal directly with the climate change treaty negotiations.

In less than two years, and in time for the 1992 Earth Summit, this negotiation committee drafted the principles of what would be the UNFCCC.[3] This environmental treaty was opened for parties' signature during the Earth Summit, and entered into force in 1994, with the goal of stabilizing GHG concentrations in the atmosphere, thus preventing dangerous anthropogenic interference with the climate system (Cherry, Hovi, and McEvoy 2014, xx). Since 1995 the UNFCCC has held annual meetings to negotiate agreements over actions toward the reduction of emissions and assess the results of the actions taken. The UNFCCC is therefore a convention treaty and, simultaneously, an international forum where currently 197 parties gather and discuss how to manage the climate.

Disagreements over the different responsibilities of each nation vis-à-vis the GHG concentrations in the atmosphere marked conversations within the UNFCCC since its very first meetings. In fact, the architecture of the negotiation process was very early on structured around a North-South divide, which was expressed in various ways: either by attributing to developed countries the responsibility for worsening the state of the planet's atmosphere—a position subscribed to by the authorities of developing countries—or by stating that developing countries could not take the same development path of industrialized nations since the global atmosphere could not sustain such a significant increase in emissions, as argued by representatives of developed countries. This position was generally considered by members of poor countries as unfair because it precluded these latter countries from achieving the same level of development as rich nations (Bodansky 1993, 479). This discussion over appropriate development paths took the form of a normative opposition between economic development and environmental conservation, presupposing that a country would always be doomed to choose between either developing itself or preserving its environment.

Although the North-South divide continues to structure the negotiations, to this day, all parties have agreed that the so-called developing countries should prioritize their economic development, leaving concerns for environmental conservation to those countries considered already developed. This decision was materialized in the first principle of the UNFCCC, which states that parties "should protect the climate system . . . on the basis of equity and in accordance with their common but differentiated responsibilities and respective capabilities" (UNFCCC 1992, 4). Although vague in terms of what exactly differentiated responsibilities and capabilities mean, the statement has been broadly understood as a recognition that industrialized countries would have to do more than the others to reduce global emissions. However, precisely because of that vagueness, the meaning of "common but differentiated responsibilities" continues to be debated (Kutney 2014, 10).

Despite this polarized negotiation dynamic, the UNFCCC has been able to achieve agreements over contentious issues such as commitments to reduce emissions and funding. The ability to forge agreements over such divisive questions lies in the way the UNFCCC agreements achieve a delicate balance between consensual statements about what needs to be done and the absence of binding actions toward such needs. This balance is reached through a carefully crafted language that alternates between concreteness and abstraction (Riles 2001, 82) or between object (the content of resulting documents) and pattern (the form of those documents). In other words, while the careful choice of words that pattern a document is abstract (and vague) enough to dodge binding commitments, it is also concrete by demonstrating that those commitments are enshrined in the final text, thus disabling eventual complaints by developing countries. For instance, instead of saying that country X will channel a certain amount of money for developing countries to help in the implementation of adaptation measures, with a concrete timetable, a final agreement would state that country X pledges a certain amount of money that will be made available for adaptation measures. The ability to reach agreements does not mean, however, that controversies have been solved. Rather, they are "papered over," which means that, through ambiguous formulations and deferrals, controversies are not inscribed in the final documents (Bodansky 1993, 493), deemed by the parties (especially parties from the Global North) as a product of consensus building.

This method of consensus building was established from the UNFCCC's inception. Throughout the numerous negotiation sessions leading to the agreement for the UNFCCC's creation, parties worked with almost illegible, heavily bracketed documents, in which literally each word had to be carefully negotiated. As words were arduously discussed and agreed upon, they would gradually fill in those brackets, until the final document was completed. According to Riles, brackets are points of potentially infinite internal expansion and/or endless internal fragmentation, meaning that while the text outside of the brackets cannot be altered, within the brackets there is potential for infinite additions (2001, 85–86). It is within that space for expansion that governments from industrialized countries have sought to dodge any type of legally enforceable language, while fragmenting key documents into recursive appendices that can only be read in reference to one other. The final result was quite indicative of what these negotiations set out to be: in the words of environmental studies scholar Gerald Kutney,

> despite all the hoopla, the nations of the world did not agree to anything meaningful in Rio [in 1992]. There was agreement to reduce emissions to 1990 levels by 2000 in principle, but these were aims with no legally binding implications. ... The only real accomplishment was that the parties agreed to keep talking and talking and ... over 20 years later, they are still talking while emissions are still rising. (Kutney 2014, 13)

Adopted by 192 parties in 1997 during the third COP, the Kyoto Protocol became the only legally binding result of the climate negotiations so far, with clear emissions-reduction goals for each developed country, however, without any legal sanctions for noncompliant parties. This meant that the legal implications of fulfilling or not fulfilling the mandatory goals for reducing emissions were exactly the same: none. Kyoto entered into force in 2005 already fraught with two situations: the United States never ratified the agreement (despite being the largest GHG emitter) and China (the second biggest emitter) was considered a developing country and therefore excused from any targets. Between 2013 and 2020, the Kyoto Protocol began a second commitment period, with even fewer countries ratifying it,[4] which explains why, despite being the only legal instrument enforcing emission reductions, signatory parties and environmentalists widely agree that "by any metric, [Kyoto] has failed to stabilize global GHG emissions" (Cherry, Hovi, and McEvoy 2014, xx).

Indeed, the means through which emissions reductions should be pursued has been a contentious topic. Part of that debate has revolved around the potential role of forests from the Global South for acting as giant storehouses of carbon (called sinks). The rationale behind this idea is that forests act as carbon sinks, precluding the release of that gas into the atmosphere. Thus, if instead of converting a forest into a farming field, or trees into timber, these forests are maintained, then, it is argued, there will be a reduction in emissions (or at least in foreseeable projected ones) through their preservation. The possibility of instituting a mechanism of compensation for avoided deforestation as part of the emissions reduction commitments, was discussed very early on within the UNFCCC, but without any agreement—for various technical and political reasons.[5] Despite the long history of disagreement over such a mechanism—known as REDD—in 2013 parties reached a consensus over its legal framework. I want to turn now to that moment, not for what it meant in terms of REDD's implementation, but because its classification as a "success" has less to do with what was achieved regarding the climate and far more with what it has accomplished in terms of expanding new forms of transnational governance.

The Breakthrough: The Warsaw Framework for REDD+

> Speaking from a personal perspective, what frustrates me in the Climate Change Convention is what is considered a success. I talked to a negotiator at the end of [the] Doha [conference in 2012] and asked him if the results for REDD were good. He said it was very good because it had created a specific mandate, to discuss, to work on a focused issue, and I, I think I even was a bit naïve, and asked him, *but wait, how can we do REDD discussing since 2005—meaning seven years!—and it is a success that the result of seven years of work is two more years of discussion?!* And then he answered, *well, you can't consider that here, because things flow on a different rhythm.* So, I think the problem is exactly that. Things flow on a different rhythm, meaning, whoever is out there, in the woods, whoever is down there in the forest, feels the problem more urgently.[6]

When I interviewed Karen in São Paulo, COP19 had not yet taken place. Given the time that, by then, REDD had been in the negotiating agenda without any concrete results, her frustration was understandable—especially because as someone advocating a marginal position, she saw herself as belonging to the group of those out there in the woods (if not literally, at least metaphorically). At the time of our conversation, Karen was a private consultant for the environmental sector, but previously she had been working with an Amazonian NGO involved in the design of the first REDD private project implemented on indigenous lands. Like many other NGO practitioners and environmentalists, she was longing for a UNFCCC agreement that would determine the rules for REDD's implementation, with the hope that such an agreement would not only kickstart REDD's implementation all over forests in the Global South, but also spur the generation of a forest carbon price high enough to ensure that forests would be more valuable standing than cut down. What parties achieved in Warsaw during COP19 in 2013 certainly fell far short of creating a global forest carbon market, nor did it spur REDD's global implementation. But what did it accomplish?

As mentioned by Karen, since at least 2005, parties at the UNFCCC had been discussing the possibility of implementing REDD as an important part of the solutions to address the problem of climate change. The gradual adherence of initially skeptical parties to an instrument that still raises so many technical questions has multiple explanations (explored in the following chapter), but the increasing propensity of policymakers to adopt market-based solutions for environmental problems must be accounted for as an important one. From the very outset, REDD was presented as a market mechanism, which means that the compensation given to developing countries for maintaining their forests standing would follow from the sale of offsets. In the same way that Kyoto's flexible mechanisms—that is, carbon markets—seduced many parties with the mantra of "markets' efficiency" and "cost-benefit analysis," REDD too promised to be a cost-efficient instrument to preserve the most important forests of the planet, thus effectively fighting climate change.

Unsurprisingly, such a market based on the forests of the Global South presents enormous logistical challenges, including accounting for the carbon contained inside the trees or how much deforestation can be allowed (timber is still an important commodity in international markets), or even how to reconcile different forest tenure regimes with the implementation of a scheme intended to be managed at a global scale. Therefore, when the UNFCCC announced that at Warsaw parties had finally agreed to the definition of the legal framework for the transnational implementation of REDD, such an achievement could only be deemed—as indeed it was—a "success" and an important "breakthrough" (UNFCCC 2013c). But how specific could this framework be regarding the procedures that countries must follow in order to implement REDD? How well could it define the sources and specific amounts of finance made available for poor countries to implement REDD? Given the diversity of forest types and deforestation drivers, how concrete could the framework be when it came to accounting carbon and addressing those drivers? Significantly, none of these specific questions were answered by the Warsaw framework and, in fact, this agreement is instead best characterized by its vagueness and polysemy. Such vagueness, however, is an important—even essential—element of the UNFCCC's language ideology.[7]

The UNFCCC's language ideology is a form of creating a world and of positioning parties in that social world through the establishment of discursive rules, acceptable terms, and the relationship between content and form. Language ideology has mostly been used to study specific groups of speakers and understand how their linguistic world influences and is shaped by their culture and social experiences. Here, I am using this theoretical framing to understand the constitution of a specific social field—that of climate policy—and how the UNFCCC affirms itself as the legitimate transnational authority within a wider epistemic community that purports to be governing the global climate.

The UNFCCC language cannot be examined outside of its field of production but rather as a constitutive part of the epistemic community that inhabits the field of climate policy (Bourdieu 1993, 33). This does not mean, however, that I am implying the existence of a single *form of talk* or ideology in these negotiations. Nor could it be such a case given the fact that this linguistic community is transnational, cutting across the international, national, and local levels. By using language ideology as an analytical tool, I demonstrate instead that during this type of negotiations, conflicting views are managed and filtered (Dent 2013) into a document that can be accepted, recognized, and subscribed to by all parties. And while conflicting views remain to be solved, the way they are managed is what allows the maintenance of a specific *social structure*, that is, the UNFCCC's power architecture.

Establishing the Rules of Performance

The Warsaw Framework for REDD+ consists of seven decisions, including a work program and guidelines for technical issues such as monitoring systems, submission of information, assessment of emission levels, and drivers of deforestation. Each decision is numbered, has a title identifying the subject, and starts with the phrase "The Conference of the Parties." Before the numbered paragraphs that account for what was decided during the conference, there are several sentences that either refer to previous decisions or state the parties' acknowledgement of certain issues (or both). Each sentence is initiated with formulations such as "Recalling," "Reaffirming," "Recognizing," or "Noting," underscoring a sense of continuity with a past moment (see dark gray circled items in figure 1).

Decision 13/CP.19*

Guidelines and procedures for the technical assessment of submissions from Parties on proposed forest reference emission levels and/or forest reference levels

The Conference of the Parties,

Reaffirming that, in the context of the provision of adequate and predictable support to developing country Parties, Parties should collectively aim to slow, halt and reverse forest cover and carbon loss, in accordance with national circumstances, consistent with the ultimate objective of the Convention, as stated in its Article 2,

Noting the urgent need for enhanced training for developing country Parties in the assessment of forest reference emission levels and/or forest reference levels,

Recalling the provisions of decisions 4/CP.15, 1/CP.16 and 12/CP.17,

Also recalling that in accordance with decision 2/CP.17, paragraphs 66 and 67, both appropriate market-based approaches and non-market-based approaches could be developed to support the results-based actions by developing country Parties referred to in decision 1/CP.16, paragraph 73,

1. *Decides* that each submission referred to in decision 12/CP.17, paragraph 13, shall be subject to a technical assessment;

2. *Recalls* that in accordance with decision 12/CP.17, developing countries may, on a voluntary basis and when deemed appropriate, submit a proposed forest reference emission level and/or forest reference level, and that such proposed forest reference emission levels and/or forest reference levels might be technically assessed in the context of results-based

Figure 1 Sentence formulations. Source: UNFCCC.

More than indicating some sort of action or orientation toward a future situation (to be regulated under the decisions made), these verbs describe a specific moment in the past that continues into the present, and that frames current decisions. The lack of new actions or movements in these verbs is not just implicit in the meaning of the verbs themselves—which point to a repetition of something that was made in the past—but also in their descriptive function. According to Austin, performative sentences cannot describe, report, or constate anything (1962, 5)—as these initial formulations do—therefore, these sentences instead of "doing something" describe a state of affairs stemming from the awareness of a past that is invoked.

This type of language is important for two reasons. First, it establishes a past that, once stabilized, cannot be changed or even contested. Accordingly, this not only presupposes a consensus over what that past is—no matter how much parties can disagree over that picture of the past—but it also forecloses possible futures. That is, by limiting the narrative of past events to a consensual narrative, the potential to consider alternative future paths to address the problem of climate change is inherently limited too. Second, this setting of the past also has implications for the participation of parties, as only those who were present in the past know exactly what is being *recalled, reaffirmed, recognized,* or *noted*. A new participant must refer to all past decisions—which frequently refer themselves to other past decisions—in order to fully understand what is being talked about in the present situation; and still, reading those past decisions will not provide any insights into the kinds of arguments held at that time. Documents on past decisions are mere filtered accounts on agreements, on consensus, not disagreements or contentiousness. This means that participants who do not possess institutional memory of these negotiations are less well positioned to participate in them. This is often the case of poor countries whose authorities cannot afford to maintain a stable team dedicated to the UNFCCC process alone. Rather, it is common to see in the delegations of poor countries members from Western environmental NGOs who volunteer their expertise to negotiate on behalf of these countries (Bulkan 2016, 95–96).[8]

The decisions follow, in numbered paragraphs, starting with verbs in the present tense. These can be classified into some sort of spectrum, from "reaffirms," "reiterates," "recalls," "recognizes," "affirms," "acknowledges," "notes," "takes note," "underlines," "invites," "encourages," "requests," "agrees," "adopts," and "decides." However, this spectrum does not mean that what could be considered the most performative verb of the repertoire—"decides"—implies a new or concrete action by any of the parties. What *deciding* actually means here is establishing the rules of the performance in the face of a *potential* action. In other words, there is a *decision* over the potential course of new actions—if they are ever taken (see light gray circled items in figure 1). None of the verbs used in the decisions commit any party to action. Instead, there is a recursive and iterating movement of reasserting the content of previous decisions, from which new agreements can be created. *However, the departure point is always a situation recognized, accepted, and therefore stabilized by all parties.*

The *recognition* repetitively affirmed in this type of document operates as a smoothing mechanism, erasing the tensions between the parties, eliding the contentious points of the discussions, and thus asserting the power hierarchies that inform these negotiations and the decisions that follow. In her examination of Australian multiculturalism, Elizabeth Povinelli argues that the act of *recognition* (as an inherent part of that multiculturalism) is simultaneously an act of formal misrecognition and a formal moment of inspection, examination, and investigation of the Other (2002, 39). According to her, the colonized subjects are expected to identify themselves not with their culture (which is a source of discomfort for those who recognize it), but with an "object of national allegiance" (39) or a "semiotically determined social space" to which the nation can relate and connect (48). This semiotic space is one in which all different indigenous specificities are flattened into a single indigenous signifier that is already situated within the limits of what can be accepted (24). Through this form of recognition, the nation can not only cohere around this accepted (and recognized) single unifier, but also feel good about such recognition. Such practices of recognition, Povinelli argues, are therefore a form of domination despite appearing as inherently good.

Although Povinelli built her argument around forms of recognition of indigenous cultures, her analysis can be applied to the logic underwriting UNFCCC's use of the verb *to recognize*. Not that the UNFCCC is dealing with cultures or identities, but it is dealing with difference: different past responsibilities, different present and future accountability, and, informing these, different political claims and goals. As such, when the documents emanating from the UNFCCC negotiating process *recognize* something, such recognition means that the multiple differences of all the parties have been flattened in a "semiotically determined" social field indexing one signifier (or a form of consensus). This means that within the social field of climate policy, there are implicit (and equally accepted) limits[9] to what can be discussed (and accepted), and therefore, the parties accept the described status quo (inscribed in the initial paragraphs of the decisions). From this follows the acceptance of a certain order, of an established hierarchy reinforcing the UNFCCC's power architecture, and of its legitimacy to govern the global climate.

Structure and Style: Recursion and the Bureaucratic Genre

According to the Oxford English Dictionary online,[10] recursive designates "a repeated procedure such that the required result at each step is defined in terms of the results of previous steps according to a particular rule." It can also designate a definition "some part of which requires application of the whole, so that its explicit interpretation requires in general many successive executions." Recursivity is an important element of the UNFCCC's language ideology, operating with multiple effects. All paragraphs in the documents are numbered, referring to decisions from previous meetings, equally numbered and codified; for example, "12/CP.16" means decision number twelve, taken during the 16th Conference of the Parties. By repeatedly referring to previous decisions, the documents not only reinforce the status quo—reminding the parties about what has been agreed upon in the past—but also render the meaning of present documents dependent on previous ones (see figure 2).

One effect of this recursivity is the creation of what I call "reading obstacles." Since reference to previous decisions is never made by its content, but instead by its code—rendering the paragraphs quite long and eventually hard to understand—one needs to know the content of previous decisions to fully understand each paragraph.

> Every time the UNFCCC takes a decision, it has to remind itself that this isn't the first time it made decisions relating to this topic, [which has] the unfortunate consequence that as time goes on, things get more and more difficult to understand. (Lang, 2013b)

Modalities for measuring, reporting and verifying

The Conference of the Parties,

Recalling decisions 2/CP.13, 4/CP.15, 1/CP.16, 2/CP.17 and 12/CP.17,

Also recalling the relevant provisions of decisions 17/CP.8 and 2/CP.17 related to the provision of support for reporting,

1. *Decides* that measuring, reporting and verifying anthropogenic forest-related emissions by sources and removals by sinks, forest carbon stocks, and forest carbon stock and forest-area changes resulting from the implementation of the activities referred to in decision 1/CP.16, paragraph 70, taking into account paragraph 71(b) and (c) of that decision, is to be consistent with the methodological guidance provided in decision 4/CP.15, and any guidance on the measurement, reporting and verification of nationally appropriate mitigation actions by developing country Parties as agreed by the Conference of the Parties, and in accordance with any future relevant decisions of the Conference of the Parties;

Figure 2 Recursivity of the UNFCCC's language. Source: UNFCCC.

These letter/number codes are thus reading obstacles because, even if one decides to skip the content of the previous decisions and just try to get the gist of the current ones, the codes constitute a form of textual hurdle that needs to be surpassed, making the reading harder, if indeed it is ever fully understandable.

As a discursive ideological device, this recursivity also establishes an important hierarchical distinction between the different participants of these negotiations by hindering knowledge of previous decisions. This process is explained by Susan Gal who argues that fractal recursions[11] (or co-constitutive distinctions) and erasures[12] (forms of eliminating those distinctions) often occur together as an ideology of differentiation by "eliding that there have been several nested contrasts made" or by highlighting similarities in the nested contrasts (2005, 27). Applying this idea to the case of the Warsaw framework, the nested decisions that appear codified in the REDD framework decisions elide, first, their content (because they are codified and not explicitly referred to by its content); second, their relationship with the current decision (be it a relationship of continuation, completion, or mere recognition of its existence); and third, the nesting process itself. That is, the recursivity of the decisions not only elides the decision process from previous COPs (including points of contentiousness), and the types of relationships between current and previous decisions, but it also imposes a *differentiation* in the drafting participants and the public of this document.

A differentiation first, between those who know the previous decisions (and can, therefore, read through the number/letter codes that refer to them) and those who do not (who are forced either to proceed with the reading without fully understanding it or to interrupt every paragraph to return to previous UNFCCC documents). This means, again, that only those who have an institutional memory of the several COPs hold the necessary holistic and comprehensive knowledge to understand what is at stake in the negotiation of each word compounding the final document. Second, there is also a differentiation between English speakers, who can easily overcome the recursion of every paragraph, and speakers of other languages who, despite their understanding of the English language, are less skilled in reading through these linguistic and numerical codes that constitute reading obstacles.

In this ideology of differentiation, then, what is being created is a "recursive public" (Kelty 2005), or an epistemic community. Chris Kelty defines a recursive public as "a group constituted by a shared, profound concern for the technical and legal conditions of possibility for their own association" (Kelty 2005, 185) or a public that conceives of its social existence through its practices and through discursive arguments. In other words, a community linked by the forms of knowledge it produces and that ensures its own reproduction as a group. In this perspective, the recursivity of the COP decisions is central to sustaining this community, whose belonging is dependent upon not just the practice of the ritual of the meeting and the normative procedures underlying it, but also the discursive assertion of sharing the knowledge and recognition of previous rituals and decisions—even if that knowledge is not evenly shared.

Another effect of recursivity is documentary coherence in both the style and content of the UNFCCC documents. According to anthropologist Iver Neumann, this type of coherent writing can be paralleled with diplomatic writing, which constitutes "an exercise in consensus building" (2012, 7). Rather than drafting something new, diplomats at these meetings are expected to reproduce what was previously known and agreed upon (7). This provides a kind of inevitability—or at least predictability—of the documents to come. It also, significantly, narrows what can be or is up for debate and establishes the narrow terms through which what is left to debate can be debated. The documents that are successively produced are, therefore, organically linked—not just through an interdependency and co-constitutive relationship of meaning, but also through a bureaucratic genre of writing. This "documentary coherence" or bureaucratic genre is ultimately a form of asserting authority and exercising power. One aspect of such authority derives from the fact that documents are vouched for by other documents and not by people (Hull 2012, 8). Since people change and rotate roles inside institutions, documents vouch for each other without depending on specific people. Hence, the authority of the negotiation process and the decisions produced is ensured by the written document, which is vouched for by a previous one, vouched for by the one before that, and so on, until reaching the limit of the depth of the recursive procedure (Kelty 2005, 206).

Moreover, since documents follow a bureaucratic genre, they do not express personal styles. In this context, documents produced by COPs are inscriptions of the UNFCCC's status quo and power architecture, expressing the UNFCCC's bureaucratic style. In seeking its institutional voice, the UNFCCC, through the repetition of "a bundle of utterances," links its parties through a single narrative, of which new instantiations—in this case, the Warsaw Framework for REDD+—confirm the previous ones and, at the same time, keep alternative narratives at bay (Neumann 2012, 56). This process can be understood as "the bureaucratic mode of knowledge production" (63), and like recognition, it imposes limits to difference, thus precluding alternative forms of conceiving climate change and solutions to it.

Bureaucracies shape political processes and constitute instruments of power (Heyman 2004, 488). Indeed, bureaucracy provides a "stable ground for authority" (Feldman 2008, 15), which demands repetition and reiteration (16). Against this backdrop, it is not surprising the number of times that verbs such as *to reiterate, to recall*, or *to reaffirm* appear in the Warsaw decisions: they constitute a form of "reiterative authority," which "relies on regularity and on an expansive view of the bureaucratic domain" (220). The repetition of previous decisions is not just about legitimizing their content in the light of the current negotiations and decisions, it is also a reiteration of the power differences between the parties, of the authority of developed countries, as inscribed in the ideology of the UNFCCC language. The "structural status quo" of these negotiations is, of course, naturalized by the language ideology (Briggs 1998, 242), while the power asymmetries shaped and reinforced by it are not explicit, as bureaucracy portrays itself as rational and efficient (Hoag 2011, 81), even and unmediated, thus as neutral as science (83). In this sense, the bureaucratic mode of knowledge production presents itself as apolitical and merely technical.

Recursion is, in sum, an important element of the language ideology of these negotiations and written outcomes and an informing element of the UNFCCC architecture. It produces multiple effects, such as documentary coherence underwriting UNFCCC's authority. This coherence—established by a consensus-based[13] bureaucratic genre—is ultimately an important source of power in the way it prevents dissension and forecloses alternative perspectives on the solutions to climate change.

Performing the English Language

The fact that UNFCCC documents are primordially crafted in English (and only later translated into other UN working languages) is another element of the language ideology under examination here, with implications at the level of participation and of participants themselves. The importance of mastering the English language in venues such as the UNFCCC is not new, since it "remains vitally important in the UN system where, despite extensive translation and interpretation services, it is still the lingua franca" (Depledge 2007, 60). However, while in the field, I became acutely aware of how language expertise can determine a country's participation in the negotiations, or their total foreclosure. During a social event with friends in Rio Branco, I met one Spanish activist who had participated in previous COPs and described to me how many of the meaningful negotiations only began after business hours, when all translators had left the meeting rooms, thus leaving delegations from poor countries unable to participate in the discussions—a practice that, she claimed, was "anti-democratic."[14] It was not the first time I had heard about these practices and it was widely known that, for instance, in the Copenhagen COP, the most important meeting from which emanated the final accord occurred in a back room between the delegations of only the United States, Brazil, South Africa, India, and China.[15]

My interlocutors in Brazil who participate in the COPs—both at federal and state level—never mentioned the dominance of English as a problem, even though Acre's state authorities pay external translation services to prepare the necessary documents to seek international certification for their carbon program and to present their REDD program in side events during COPs or other international venues. But in Mozambique, the lack of English skills among the governmental staff is a serious issue. One of my interlocutors, Natércia, a high-ranking official inside the Ministry of Agriculture, regretted the fact that the World Bank (as the institution leading REDD in the country) does not provide enough slots to learn English, identifying language as the most pressing obstacle to Mozambique's participation in the UNFCCC negotiations.[16] Mário, another official from that ministry, was blunter in describing the problem:

> Language is the biggest difficulty for a country like Mozambique in these negotiations; when they [officers from the Mozambican Ministries] arrive at the COP and finally start to understand the content of the several proposals, developed countries are already making changes to those proposals. These countries go to the meetings very well prepared; Mozambique goes to watch and eventually learn something. A big investment in training is needed to enable Mozambique's participation.[17]

Adding to the dominance of English, countries like Mozambique also lack the institutional knowledge that can only be built over time and through a consistent and continuous participation in this type of conference. This specific knowledge is not only at the heart of bureaucratic power (Weber [1946] 1973)—the type of power deployed by the UNFCCC—but is also part of a meaning-making process that shapes social reality and informs action upon that reality (Barnett and Finnemore 2004, 29–30); that is, without being in possession of this specific knowledge, Mozambican negotiators cannot even suggest courses of action (within the limits of what the UNFCCC considers acceptable or recognizable),[18] let alone contest what is being suggested by other parties. Mozambican officials have been "learning by doing," stated Natércia, because they have not been trained in climate change issues and therefore lack the knowledge of previous negotiations and their outcomes.

Despite the problems of some of my interlocutors in mastering the English language, many of them—especially the more linguistically skillful—deploy many of the English terms and acronyms of the UNFCCC jargon when talking in Portuguese. During my interviews with Brazilian and Mozambican officials, as well as NGO staffers, I was frequently presented with English words such as "stakeholders," "landscape," "draft," "readiness," "concept note," "reference levels," "framework," "capacity building," "guidance," "leakage," or with untranslated acronyms like LDCs, REM, R-PIN, and RPP, among others.[19] This form of code-switching can embody a wide variety of different language contact phenomena, with various functions (Heller 1988, 2), especially if each communicative event and respective interlocutors are situated within a broader political economic and historical context (Gal 1988, 247).[20] However, and despite all the contextual differences among my interlocutors in Acre, Brasília, and Maputo, their usage of English words can be interpreted as expressing two distinct situations.

On the interactional level, these words were used to define a community and their roles within it—especially in the case of the acronyms—thus establishing a boundary between those who understand the meaning of these words, and therefore can be part of the conversation, and those who do not (Heller 1988, 1). In this sense, their usage of English words was both a statement about their belonging to this international climate community (and of their role in it) and a form of assessing my own position regarding the boundaries of that community (did I know the meaning of the acronyms? Was I a specialist in the topics discussed at the UNFCCC under this language?). The punctuation of interactions in Portuguese with English words constituted exchange moments during which my interlocutors negotiated a particular identity in relation to me (Myers-Scotton 1993, 152). Moreover, while my interlocutors asserted their belonging to this community, at the symbolic level their code-switching was also a stylistic choice in their discursive performance, commenting on their perceptions of themselves and their relations with others and invoking the authority emanated by the UNFCCC language (111).

On a broader level, though, the use of English words during conversations held in Portuguese can also be understood as a form of subject formation. That is, by participating in the UNFCCC meetings, by reading the documents produced there, and by attendees talking to their peers in that venue, the language used is necessarily indexed to a particular social field—that of the UNFCCC, its architecture, and its role governing the global climate (Hanks 2000). This language, therefore, starts to shape the way in which these participants see and understand the world (Englund 2006, 29) or, in this case, see and understand the problem of climate change and its potential solutions.

In the end, setting aside common claims about the hegemonic role that the English language plays in transnational organizations like the UN, there are three other important aspects to emphasize in the performance of the English language. The first one is at the level of participation: since poorer countries do not master the English language, nor do they have the capacity to maintain a dedicated team to the UNFCCC process, their full participation is necessarily curtailed. Notwithstanding this mechanism of differentiation, Portuguese speakers deploy English terms in conversations about climate change; hence, the other two effects, at the level of participants themselves. One is the community-building and boundary-making around the use of a particular language marked with acronyms and a specific jargon; that is, the performativity of English language is a constitutive element of the climate epistemic community. Second, there is a subjectification effect operated by language, which not only informs the perspectives over the problem of climate change and the possible solutions to it, but also, and concomitantly, flattens difference—both the difference indexed to a language other than English and the difference in potential alternative narratives about climate change.

Deciding What Decisions Might Look Like

The Warsaw Framework for REDD+ represents an exercise of bureaucratic knowledge production and satisfies stylistic conventions following an aesthetic based on repetition and a "properly patterned language" (Riles 2001, 80), of which the tenses of the verbs in the beginning of each paragraph constitute a fine example. For these reasons, the language of the decisions does not point to anything particularly new, nor concrete, regarding countries' needed actions to implement REDD. Yet this legal framework was considered a *breakthrough*. If nothing particularly new or concrete was added to previous decisions, why was this negotiation deemed a success by the UNFCCC?

The seven decisions made in Warsaw concern the elements necessary to implement REDD: funding and who will pay for it; monitorization of forests; safeguards; reference levels of emissions (that is, a base from where emissions are counted); measuring, reporting, and verification of forest emission levels; and drivers of deforestation. Each one of these themes provide significant examples of how the language used in each decision endows REDD with the potential to become real, promoting the subscription of parties, yet without committing them to any of the decisions.

FINANCING

The first decision provides all the technical and methodological details on how developing countries should seek funding for REDD activities, including the kind of information to be provided and how to provide it. It mentions that funding can come from several sources but does not specify any of those sources, nor does it discuss the amounts available and needed to carry out REDD activities. Although the need for "adequate and predictable results-based finance" is recognized (UNFCCC 2014a, 24), it is not explained how *predictability* is to be achieved. Developed countries are, however, *encouraged* "to provide financial resources." Worth underscoring here is that the verbal forms "agrees" and "decides" are used in relation to information issues only: what is *agreed* on is the type of information that countries seeking finance need to provide; what is *decided* is the creation of an information hub and what information it will contain—including the *agreement* that such information will be provided in consultation with the "developing country Party concerned" (UNFCCC 2014a, 25).

Another detail worth emphasizing is the fact that none of these paragraphs are mandatory, which is visible in the number of times expressions such as "as appropriate" or "in accordance with national circumstances" appear (28). This lack of legal force is often criticized by environmentalists who are concerned with the possibility of governments using "national circumstances" to justify continuing deforestation activities. The website redd-monitor.org exemplifies this concern when it poses the question:

> How does the Warsaw REDD deal address the danger that this provides a get out of jail free card, allowing, say, Indonesia to continue deforesting, because of the national circumstances of large coal deposits below the forests of East Kalimantan? (Lang 2013a)

For those advocating for the implementation of REDD, the striking point about the decisions on funding is that there is not a single line stating how much will be made available, by whom, or who can access it. As such, the Warsaw framework "is not an agreement by anyone to finance REDD. Rather, it's an agreement on what REDD finance might look like if there was any finance" (Lang 2013b). In another example, during a public event about the Warsaw COP, held in Washington, D.C., the director of a company working on carbon projects complained about the Warsaw decisions being not about real money but about choosing national entities who will assess the results of REDD activities, so they can be paid, concluding, "There are no commitments for money!"[21] However, these decisions on funding suggest the potentiality of payment for REDD; that is, they allow imagining REDD being implemented because it will be funded. And the possibility of imagining such a picture suggests that REDD can become a reality.

FOREST MONITORING SYSTEMS

The emphasis on national circumstances and the voluntary tone of the whole Warsaw Framework for REDD+ is again underscored in the decision on "national forest monitoring systems" in which the expressions "as appropriate" or "if appropriate" appear six times in five short paragraphs (UNFCCC 2014a, 31–32). It states that national monitoring systems "should take into account" previous decisions and the guidelines of the IPCC—the emphasis being in the form *should*, which is repeated to describe technical issues of these systems. And it states that national forest monitoring systems *should* "enable the assessment of different types of forest in the country, including natural forest, as defined by the Party" (UNFCCC 2014a, 31), which, according to some environmental NGOs, opens the possibility for tree plantations to be included in the REDD mechanism (see figure 3).[22]

4. *Further decides* that national forest monitoring systems, with, if appropriate, subnational monitoring and reporting as an interim measure as referred to in decision 1/CP.16, paragraph 71(c), and in decision 4/CP.15, paragraph 1(d) should:

    (a)   Build upon existing systems, as appropriate;

    (b)   <u>Enable the assessment of different types of forest</u> in the country, including natural forest, <u>as defined by the Party</u>;

    (c)   Be flexible and allow for improvement;

    (d)   Reflect, as appropriate, the phased approach as referred to in decision 1/CP.16, paragraphs 73 and 74;

Figure 3 Forest monitoring systems. Source: UNFCCC.

The vagueness of this decision regarding what supposedly is concrete and consensual—a forest—can be explained as a form of broadening the scope of REDD's application, an important step toward making REDD real. By broadening the scope of REDD's application, it is possible to include a larger number of activities implemented under the REDD template while also promoting the enrollment of a larger group of people (see Mosse 2005).

INFORMATION ON SAFEGUARDS

Forms of intervention by transnational organizations like the UN or the World Bank—be they developmental, environmental, or otherwise—have come to increasingly include measures that purport to prevent (or mitigate) the unwanted or harmful impacts of such interventions upon groups of people (notably indigenous) and the environment. These precautionary measures are called safeguards. In certain domains of intervention these safeguards can be sufficiently strict to prevent these works from happening altogether due to their anticipated negative impacts. However, the Warsaw framework maintains a decidedly voluntaristic tone regarding such safeguards. Therefore, on the decision on "the timing and the frequency of presentations of the summary of information on how all the safeguards . . . are being addressed and respected," it is *agreed* that this information "could be provided, on a *voluntary* basis, via the web platform" (UNFCCC 2014a, 33).[23] It is also *decided* that "developing country Parties should start providing the summary of information" and that such information "should be consistent with the provisions for submissions . . . and, on a voluntary basis, via the web platform" (33). Again, what is decided is limited to the voluntary provision of information.

FOREST REFERENCE LEVELS

In the decision regarding the "Guidelines and procedures for the technical assessment of submission from Parties on proposed forest reference emission levels and/or forest reference levels," the annex providing those guidelines stands at first glance as being far more concrete. Even though this annex (composed of eighteen paragraphs) is supposed to be eminently technical, the politics of these negotiations continue to lurk just beneath the discursive surface, being most explicit when it is mentioned that one of the objectives of the technical assessment is to "offer a facilitative, non-intrusive, technical exchange of information" (UNFCCC 2014a, 36); and also later in the document, when it is declared that the "assessment team shall refrain from making any judgement on domestic policies taken into account in the construction of forest reference emission levels" (37). This is an important statement because first, it uses the form "shall" instead of "should," imprinting some degree of enforcement into the decision; and second, because "domestic policies" are precisely the kinds of issues that can be at the origin of deforestation or that can be used to tweak the reference levels provided by each country. It should be noted, then, that the most concrete decision in the framework thus far not only presents itself as "technical," but also makes the imperative—an order—of avoiding any kind of interference in the politics of each country. Of note too is the fact that such a contentious issue—defining a baseline from which emissions are counted—is rendered technical (Li 2007), thus voiding further (political) discussions on the matter.[24]

Another important element of this annex is its insistence on the "consistency," "transparency," and "accuracy" of the information and data provided—three polysemic themes that are connected and reinforce one another. I will return to these three themes since they inform the whole UNFCCC architecture.

MEASURING, REPORTING, AND VERIFYING

The decision on the modalities for measuring, reporting, and verifying (a set of three activities concentrated in the acronym MRV), with a total of fifteen paragraphs, is by far the one with the most *decisions* (eight) and *agreements* (three). This can be explained by the fact that the metrics of emissions, including their reporting and subsequent verification by a third party, are considered the most technical issues addressed. Despite being presented as eminently technical; these questions are highly contentious. In fact, it is precisely its potential as a source of conflict that explains why the language of this decision turns it into an exercise on smoothing such contentiousness. To this end, it is replete with expressions such as "is voluntary," "may interact," "may seek," "should provide," "in accordance with national circumstances," or "taking into account national capabilities" (UNFCCC 2014a, 39–40), implicitly and explicitly denoting unconstrained resolutions that do not commit any party nor interfere with national sovereignty. Although the obligations to provide *accurate* information are inscribed in the text, such obligations are also *voluntary* and provided according to the circumstances of each country. Finally, most paragraphs on information or data contain the adjectives "transparent" and "consistent," emphasizing the idea of trustworthiness between parties. This voluntaristic tone and pointing to transparency are due to the fact that the MRV of carbon emissions has a great potential for conflict and mutual accusations.

DRIVERS OF DEFORESTATION

Finally, the last decision, which addresses the drivers of deforestation and forest degradation, is one of the shortest (only five paragraphs), despite its supposed centrality in the wider context of policies to fight deforestation. If the rest of the framework is already far from specific, this portion of the document is particularly vague in "noting the complexity of the problem" or the "many causes" of deforestation, *encouraging* all parties "to continue their work to address drivers of deforestation" and "developing country Parties to take note of the information from ongoing and existing work on addressing the drivers of deforestation" (UNFCCC 2014a, 43). Unlike other decisions, this one does not have a single paragraph initiated with the verbal forms "agrees" or "decides." Rather, mere *notes* and *encouragements* populate the section along with "reaffirms" and "recognizes." The politics of this decision do not limit themselves to the well-crafted balance between vagueness and concreteness, nor between content and form. The content itself reveals some of the problematic assumptions that inform the rationale for mechanisms such as REDD to be championed as the solution to the problem of climate change—notably, the idea that deforestation in the Global South is unaddressed by lack of capacity (or even unwillingness) by authorities of developing countries.

In the opening statements, it is mentioned that "livelihoods may be dependent on activities related to drivers of deforestation and forest degradation," and although it is *noted* that "addressing these drivers may have an economic cost and implications for domestic resources," the parties also *recognize* that "actions to address these drivers are unique to countries' national circumstances, capacities and capabilities" (UNFCCC 2014a, 43). There are no mentions of international deforestation drivers, such as the demand for timber (as is the case of the Chinese thirst for Mozambique's forests) or the increased consumption of beef (the main drivers of Amazonia's deforestation), which means that only local and national "livelihoods" are to be held accountable for deforestation (see figure 4). This is a discourse with great currency among environmental organizations, especially those involved in the implementation of REDD, which, tellingly, never addresses global causes of environmental degradation. This logic echoes similar situations of development interventions, like the one analyzed by James Ferguson (1994) in Lesotho, in which the World Bank's diagnosis for the country's problems ignored its structural ramifications to world markets or even to South Africa, restricting its intervention within the scope of Lesotho's national economy and political body. In other words, instead of considering all the international drivers for deforestation, the destruction of forests in the Global South is isolated from its broader context and explained solely by local or national factors.

By limiting the problem of deforestation to the national scale, the decision not only elides the responsibilities of developed countries in deforestation, but also shifts that responsibility to countries from the Global South. Nothing is said about the responsibilities of developed nations in the deforestation of poor countries due their high demands for beef, timber, or fossil fuels, nor is anything said about deforestation in industrialized countries. Rather, it is implied that deforestation occurs in developing countries only, which is an inherent assumption in the whole REDD architecture—REDD was designed to be implemented in the Global South, and deforestation is a problem in developing countries, which ought to be solved with the help of industrialized nations.

The aforementioned spatial disjuncture of the UNFCCC's narrative is apparent, although never acknowledged in the decisions. In this narrative, climate change is a global problem, involving the unbounded global atmosphere. Yet salvation relies on the forests of only half of the planet—that is,

Decision 15/CP.19*

Addressing the drivers of deforestation and forest degradation

The Conference of the Parties,

Recalling decisions 2/CP.13, 1/CP.16 and 2/CP.17,

Noting the complexity of the problem, different national circumstances and the multiple drivers of deforestation and forest degradation,

Also noting that livelihoods may be dependent on activities related to drivers of deforestation and forest degradation and that addressing these drivers may have an economic cost and implications for domestic resources,

1. *Reaffirms* the importance of addressing drivers of deforestation and forest degradation in the context of the development and implementation of national strategies and action plans by developing country Parties, as referred to in decision 1/CP.16, paragraphs 72 and 76;

2. *Recognizes* that drivers of deforestation and forest degradation have many causes, and that actions to address these drivers are unique to countries' national circumstances, capacities and capabilities;

3. *Encourages* Parties, organizations and the private sector to take action to reduce the drivers of deforestation and forest degradation;

4. *Also encourages* all Parties, relevant organizations, and the private sector and other stakeholders, to continue their work to address drivers of deforestation and forest degradation and to share the results of their work on this matter, including via the web platform on the UNFCCC website;[1]

5. *Further encourages* developing country Parties to take note of the information from ongoing and existing work on addressing the drivers of deforestation and forest degradation by developing country Parties and relevant organizations and stakeholders.

Figure 4 Drivers of deforestation. Source: UNFCC.

by saving forests in the Global South, climate change can be avoided. The drivers of deforestation are, however, national, and the solutions to those drivers are locally implemented. Not only is there a spatial tapering of the problem—where the dimensions of the solutions mismatch the dimensions of the problem—but developed countries remove themselves from the whole problem as well. Instead, industrialized nations reposition themselves as the agents of global salvation by diagnosing the problem, its scope, and its solution, and by presenting the methods for the application of such a solution.

Although this narrative on climate change and forests has been enshrined in the Warsaw framework, that does not necessarily mean that developing countries accept it altogether. The disagreements are simply elided from the UNFCCC's final documents due to the bureaucratic genre that informs them. Smoothing the language or emphasizing a voluntaristic tone are forms of eliding tensions and potential disagreements. Similarly, the deflection of industrialized countries' responsibilities in global deforestation, as well as their systematic dodging of financing commitments to help forest conservation, have also been set aside in the drafting of the Warsaw agreement but have increased the lack of trust between negotiating parties. This lack of trust is particularly apparent when, for instance, industrialized countries continue to demand more transparency and reliability of the data about forests of the Global South—implicitly stating that developing countries are presenting unreliable data. For developing countries' delegations, such claims are a demonstration of prejudice, and a delaying tactic to avoid the disbursement of funds. Ultimately, this climate of suspicion transpired into the vocabulary of the framework in a reversed form.

Trust and Transparency: Negotiating Suspicions

Looking at the entire framework, *accuracy*, *consistency*, and *transparency* emerge as rhetorical themes (Hill 1998) of the UNFCCC language. *Accuracy* and *consistency* appear mostly in the methodological parts of the documents, not only with the efforts to format the information to be submitted according to similar stylistic characteristics, but also in the standardization implemented by the guidelines. These efforts toward standardization correspond to a depoliticization effect in the UNFCCC language ideology.

In these methodological moments, there is an effort to "scientize" and "technicize" the language used, which is visible not only in the instructions toward uniformization, but also in the procedures for the composition of teams of experts, whose skills ought to be vouched for by guidelines from previous decisions (recursivity all over again) and by other UNFCCC experts. The submissions of emissions levels by the parties are to be *technically*[25] assessed, suggesting a certain neutrality to this process. It is precisely in one of the most politically fraught issues of this negotiation—emissions accounting—that the language becomes more technical and supposedly neutral. Financing is another matter in which the technical language dominates. However, whereas the political load of financing is dealt with through vagueness[26] (regarding funding sources and their predictability), the question of emissions accounting, baselines, and reference levels is neutralized through a language of metrics.

The problem of emissions accounting is that, on one hand, the national submissions of emissions can be manipulated by each party to claim more financial benefits (as in, *my country reduced deforestation more, therefore deserves more money*), or to conceal higher deforestation levels. For instance, during the event I attended in Washington, D.C., about the Warsaw COP, one participant mentioned that the decisions on the reference levels were dubious, since only Indonesia and Brazil have strong data on their historical emissions that can provide a strong baseline. Adding that "most of these countries [i.e., poor countries with forests] do not have any data on their forests, which compromises the environmental integrity of the agreement."[27] This concern about reference levels and baselines is tied to what some perceive as an impossible standardization given forests' diversity and the different capabilities of each country. From which follows an ineluctable suspicion regarding the information provided by these countries' authorities on their emissions levels.

Higher emission-reduction levels also mean more money that industrialized countries do not want to pay, while assessing emission levels by a third party could require a level of interference in nations' internal affairs that would render an agreement impossible. The context of suspicion that enfolds these negotiations is thus addressed by the themes of *accuracy* and *consistency* that, through standardization, defuse political tensions and normalize the relations between countries. Here, it is important to recall James Scott in order to understand the political implications of standardization—as part of a high-modernist ideology—in its efforts to render a certain social arrangement legible, appropriable, and controllable (1998, 219). In the language of the UNFCCC, the ideological tropes of science and *techne* are the means to control the arrangement of the parties, defusing the politically fraught untrustworthiness that marks these negotiations. *Transparency* is, likewise, part of this rhetoric of normalization and political neutrality. Yet, in a context marked by distrust, discourses of transparency only come to reinforce the atmosphere of suspicion.

Transparency informs the negotiation architecture itself,[28] being constantly upheld as the necessary condition for a successful post-Kyoto Protocol agreement. The visual idea of transparency is a recurrent trope in

2. The technical assessment of the data, methodologies, and procedures used by the developing country Party under assessment in the construction of its forest reference emission level and/or forest reference level in accordance with decision 12/CP.17, chapter II, and its annex, will assess the following:

(a) The extent to which the forest reference emission level and/or forest reference level maintains consistency with corresponding anthropogenic forest-related greenhouse gas emissions by sources and removals by sinks as contained in the national greenhouse gas inventories;

(b) How historical data have been taken into account in the establishment of the forest reference emission level and/or forest reference level;

(c) The extent to which the information provided was transparent, complete,[1] consistent and accurate, including methodological information, description of data sets, approaches, methods, models, if applicable, and assumptions used and whether the forest reference emission levels and/or forest reference levels are national or cover less than the entire national territory of forest area;

(d) Whether a description of relevant policies and plans has been provided, as appropriate;

(e) If applicable, whether descriptions of changes to previously submitted forest reference emission levels and/or forest reference levels have been provided, taking into account the stepwise approach;[2]

11. *Further decides* that, as part of the technical analysis referred to in decision 2/CP.17, annex IV, paragraph 4, the technical team of experts shall analyse the extent to which:

(a) There is consistency in methodologies, definitions, comprehensiveness and the information provided between the assessed reference level and the results of the implementation of the activities referred to in decision 1/CP.16, paragraph 70;

(b) The data and information provided in the technical annex is transparent, consistent, complete[2] and accurate;

(c) The data and information provided in the technical annex is consistent with the guidelines referred to in paragraph 9 above;

(d) The results are accurate, to the extent possible;

Figure 5 Transparency, accuracy, and consistency. Source: UNFCCC.

contemporary ideologies of language in Western nations (Sanders and West 2003). Deemed as a superior goal in communication, transparency works as part of a strategy to elicit increasingly specific information within a vague framework (Brenneis 2006, 43–44). The decisions inscribed in the Warsaw framework constitute a great example of such a vague arrangement that still forces parties to deliver very specific information regarding their forests, economic activities, and legal systems. Genealogically connected with Enlightenment principles of reason and truth, "transparency claims constitute yet another way of celebrating the rationality of modern society" (Sanders and West 2003, 7). Transparency denies suspicion or doubts, almost declaring trust by fiat. Indeed, the lack of trust between parties is constantly referred to as the main cause for the slowness of the negotiations, or even their failure. And while transparency is hailed as a goal, it is not reflected in the opaqueness of the decisions. Although claiming transparency, the Warsaw framework aptly demonstrates Sanders and West's assertion that "modernity, paradoxically, generates the very opacities of power that it claims to obviate" (2003, 16). That is, the Warsaw decisions are rendered opaque in form (through recursivity and what I called "reading obstacles") and in content (through vagueness and not binding decisions).

The ideological formation of the UNFCCC's language is not just revealed through the transparency trope. The lack of trust between parties also implicitly frames the negotiations, namely through the acronym MRV, which stands for measuring, reporting, and verification. This acronym is present throughout the grey literature on climate change and constitutes a symbol[29] that, for the climate policy community, seems to have referential meaning only; that is, ideally, it does not depend on a context to have meaning—MRV is simply something inherent to carbon accounting. One of the characteristics of the symbol, as defined by Charles Sanders Peirce, is that it is removable from context. Here, I say *ideally* because who performs the MRV and where it is performed can be a source of suspicion. Hence, the *who*, and the *where* can be the context that contradicts the (idealized) exclusive referential meaning of MRV, giving it an indexing feature (Hanks 2000).

The three terms are not separable nor arbitrarily ordered, forming a single expression that points to something inevitable, whose need is unquestionable, while also being necessary to ensure transparency, consistency, and accuracy. In this sense, it has a directive function (Hanks 2000, 124). Something that has been through the MRV process is thus transparent, consistent, and therefore, trustworthy. Trust is directly dependent upon performing MRV. Although MRV has emerged as part of the methodological aspects of reducing carbon emissions, currently, it is used for everything to be implemented within the UNFCCC context, including funding,[30] or even performing MRV on the MRV. For instance, when I interviewed the spokesperson for the Brazilian Ministry of the Environment, she expressed her frustration toward the requirements that were being demanded by developed countries regarding emissions levels, which were tantamount to implementing an MRV process over the already existent MRV. These demands were, in her perspective, outrageous given these countries' double accounting and rebranding of aid money.

The MRV acronym is not deconstructed in any of its three parts, nor is it problematized. Rather, it is accepted in its whole, as a symbol. But if something is measured and reported (according to standardized methods), why does it need to be verified? Here lies the indexical meaning of the symbol MRV: verification is the consequence of the presumption that things can be concealed while being measured, and manipulated in reporting, especially if measured and reported by developing countries. From which it follows that developed countries are essentially forced to verify the veracity of what has been reported. This need for verification also expresses the ideology of audit, which emerges as one core trait of neoliberalism (see Strathern 2000; Shore and Wright 2000). There cannot be a measurement or a report without a verification, which is operated by a *neutral* party,[31] ensuring *consistency, accuracy*, and *transparency*. In this context, MRV is the deictic counterpart of transparency.

In the same way that disagreements, tensions, and contentiousness are elided from the UNFCCC's narrative, the ideology of the language used by the UNFCCC hides mistrusts and suspicions, by asserting the value of transparency, consistency, and accuracy, both as goals and achievements. These processes of elision, flattening, and depoliticizing that I have been describing throughout the chapter cut across the format and content of the Warsaw Framework for REDD+ and structure the language ideology of the UNFCCC. Thus, if the Warsaw agreement—as the documentary expression of the UNFCCC's language ideology—flattens and even erases the complaints and demands of an entire bloc of parties (the so-called developing countries), how could it still be deemed a breakthrough and a success? One could also wonder: success to whom? It is time, then, to analyze what success means.

"A Clear Breakthrough for Action on Climate Change"[32]

Not everyone agrees with the characterization of the Warsaw COP regarding the REDD platform as a success. Many NGOs consider the agreement weak and worrisome given the lack of any funding commitments or legal force. Among NGOs there are mostly two types of critiques to Warsaw: one that points to the lack of concrete results in terms of binding policies and thus of "real" action against climate change; and another unsatisfied with the absence of a clear path toward a global market for forest carbon. Regarding the latter, the director of carbon projects in a US-based nonprofit working in several countries and responsible for several REDD projects in Acre stated the following:

> Overall, I think Warsaw was a step in the right direction for REDD+. A lot of technical decisions were made which is great. Yet, there is still no compliance market that accepts REDD+ projects which remains a significant barrier.[33]

Other representatives from critical NGOs continued to express their frustrations in the aftermath of Warsaw, as I was able to realize during interviews in some of the D.C.-based NGOs involved in REDD projects and carbon finance. A representative from a US NGO working on development, agriculture, and natural resources expressed his skepticism regarding the negotiations:

> I think the negotiations are in trouble. I don't see any pathway forward. You know, they'll continue to have meetings. And meetings have value, these exchanges take place, and knowledge is gained. But I don't see the market value coming. I think it's going to lead to frustration.[34]

The director of an American-European consulting company for projects to reduce carbon emissions was more outspoken about his disbelief in the value of such a negotiation process:

> I'm so tired of governments' inertia, especially during the negotiations, that I just want to join companies and other organizations and DO something. To governments, there is no sense of urgency whatsoever. And even when they can agree on something, it is such a light agreement, that does not solve any problem. To me, Warsaw was lame.[35]

At the same time, the UNFCCC has been increasingly challenged over its role as a platform for negotiations that are more and more about financing issues. The discrediting of the UNFCCC as a negotiation platform has surely been denounced by some NGOs, especially those that operate closer to carbon markets. This issue was raised several times during the event I attended in Washington, D.C., about the Warsaw meeting.[36] The reason claimed for taking the negotiations out of the UNFCCC is the fact that these are increasingly negotiations about money; therefore, they would need a finance platform.[37] Against this backdrop, and given both environmentalists' criticism and the mild reception of the REDD outcomes by both developed and developing countries' governments, one wonders why the UNFCCC is hailing this process a "clear breakthrough."

The most popular narrative explains the success of the REDD negotiations in contrast with the lack of results in the rest of the subjects being discussed in the meeting, and with all the drama that punctuated the COP. The adjective successful was used by the director of one environmental company during the D.C. event I have been referring to. In her intervention, she claimed that the REDD platform during the Warsaw COP was hailed as a success because the decisions were made before the drama that marked the final part of the COP.[38] Indeed, except for the Copenhagen COP, whose failure and "ghosts"[39] (of secret documents,[40] blackmail,[41] and US espionage over the delegations of certain countries[42]) still haunt the UNFCCC negotiations, the Warsaw meeting was probably the richest in off-script events.

To begin with, the COP started under the shadow of the tragedy of the Philippines typhoon, which, for environmentalists and governments from poor countries, constituted a clear demonstration of the urgency to address climate change. Thus, given the stalling of the negotiation process, the head of the Philippines delegation announced at the opening of the conference that he would fast until progress was achieved. At the same time, adding to the political tensions, the hosting country decided to organize a parallel convention on coal. This enraged environmentalists and was perceived as a provocation by the Polish authorities, who have voted against almost every EU decision to address climate change. Despite the lack of expectations for significant advances during the negotiations, in the final days, the frustration was mounting, and in an unprecedented move, NGOs, social movements, and unions walked out of the COP, protesting the parties' lack of ambition and the unlikeliness that they would achieve effective results (Vidal and Harvey 2013). Finally, to make matters worse regarding Poland's commitment to these negotiations, the Polish environment minister, who was presiding over the event, was fired by Prime Minister Donald Tusk, allegedly displaced by his minister's slowness in approving shale gas projects. The outcome of this drama was a general fallback in legal obligations, with the replacement of reducing emissions commitments by reducing emissions contributions, and a retreat from previously agreed-upon emission-reduction goals by Japan and other industrialized countries. In short, while all the other Warsaw COP documents reflected this drama by regressing on the achievements of previous negotiations, the Warsaw Framework for REDD+ emerged as an actual progress moving negotiations forward.

Refusing to acknowledge what is considered by several NGOs to be a total failure,[43] the UNFCCC press release after the event stressed the fact that the conference ended by "keeping governments on a track towards a universal climate agreement in 2015 and including significant new decisions that will cut emissions from deforestation and on loss and damage" (UNFCCC 2013b). This narrative is therefore based on the overall very low expectations about results—thus, anything considered as tantamount to an agreement can, indeed, be considered a success.

Notwithstanding, other explanations can be offered to explain the hailed success of the Warsaw COP. David Mosse argues that the construction of *success* in development projects is about establishing a compelling interpretation of events and enrolling a network of supporters, linking them to the success of such projects (2005, 158–59). Ultimately, a project is successful not because it turns design into reality, but because it sustains policy models offering a significant interpretation of events (185). This interpretation of what success means can be exemplified in a critical perspective of the Warsaw results:

> There was a strong willingness to agree on any kind of text, including on issues like MRV, reference levels and safeguard information systems, where countries are more or less free to provide any kind of information according to any kind of system they want [because that will maintain REDD as a mechanism that does not address the drivers of deforestation; and if] . . . drivers remain unaddressed at the demand side, this inflates the baseline, and makes it possible for countries to demand far more funding to reduce forest loss. (Lovera 2013)

According to this type of critique, the REDD framework does not address the drivers of deforestation but allows countries like Brazil to claim for additional funding, because the country is able to show a reduction in its deforestation levels. In other words, the technical decisions taken in Warsaw—those that, according to my argument, diffused political contentiousness through a language of standardization and metrics—are vague enough for countries in the Global South to use them in their benefit, by claiming better results in deforestation reductions even in the face of lack of funding commitments by industrialized countries.

Additionally, those more adamant against what was being required from poor countries regarding MRV were convinced to agree: after derailing the REDD negotiations in 2012, the Brazilian authorities became the program's main champion in 2013, which may be explained by the large financial contribution that Norway channeled to the Amazon Fund that same year.[44] In this context, events can effectively be interpreted as being successful: on one hand, developed countries are not legally required to provide funding for REDD (even though some money had to be spent to convince important parties, like Brazil, to accept an agreement), while their responsibilities for both climate change and deforestation are erased from the documents. And on the other hand, developing countries are provided with a legal framework that offers the potential for claiming funds in the future—if they are made available—while their sovereignty over their forests is not openly challenged by any MRV system. In parallel, the UNFCCC can claim advances in the negotiation process because an agreement between 195 parties was reached. That is, the REDD negotiations were a breakthrough not because they will enable this framework to be translated into new and effective policies and activities (after all, there is no money to implement any of the preparatory measures,[45] let alone REDD activities) but because they allow for the persistence of this policy model by maintaining parties enrolled in the UNFCCC narrative.

The language used in the REDD decisions can thus be seen as part of an effort to engage supporters to sustain a narrative of success (Mosse 2005). Parties are *invited* or *encouraged*—not committed—to do certain things (mostly, to continue doing what they are already doing). Information to sustain policies and actions needs to be provided, but on a voluntary basis. The decisions are vague, but they open the possibility of imagining, enabling the potential for future actions that meet everybody's expectations: there might be money available if deforestation levels are reduced; poor countries might be able to claim for those funds if they provide the right numbers for reduced deforestation levels; REDD might became a global forest carbon market if parties continue to be engaged in furthering the mechanism; there are significant opportunities for investment in the Global South if REDD is maintained in the negotiating table; REDD might constitute the salvation for the planet if more progress is achieved; and so on. In practice, this means that REDD activities already in implementation under private initiative and oriented toward the private markets will not be affected and can actually benefit from the perspective of future advancements in the following REDD negotiations. On the policy side, this means that whatever is already under implementation, like pilot projects, may continue to proceed, and if there is money available, more REDD activities can be implemented—as long as they follow the standardized methods and *transparency* guidelines agreed upon in Warsaw.

The role of the ideological language used by the UNFCCC is also successfully accomplished in sustaining and reinforcing its recursive epistemic community by providing the needed (repeated) points of reference in the history, evolution, and experience of this community (Kelty 2005, 193–94)—namely, through technical terms like "baseline" or "reference levels," the symbol MRV, and the tropes of *transparency*, *consistency*, and *accuracy*. The success of the UNFCCC language ideology was further reinforced through the performativity of the ritual (Silverstein 1998, 138) of the COP meeting, in which the parties *recognized*, once again, the structural status quo inscribed in the recursivity of previous decisions. In what concerns the UNFCCC's architecture, the Warsaw Framework for REDD+ cannot be deemed anything other than successful, as the language ideology sustained the social arrangement of the parties, their continuous enrollment, and the negotiation model, enabling its reproduction through a deeply layered recursive process.

Conclusion: Enrolling Transnational Governance

In this chapter, I analyzed the seeming paradox of having an agreement classified as a success that, however, does not define concrete actions or establish commitments by the signatory parties. Considering the history of the UNFCCC negotiations, having all parties agreeing on anything—even if it seems an inconsequential document—may indeed be a success, given the profound dissensions on matters such as cumulative responsibilities, compliance on actions to reduce emissions, and funding. In its almost thirty years of negotiations, the UNFCCC parties only agreed to one legally binding agreement—the Kyoto Protocol—and the one that replaced it—the Paris

Accord—is merely based on voluntary contributions. Despite the vague and apparently fruitless results achieved so far, the UNFCCC process should not be mistaken for a mere ritualistic event meant to keep things the same.

The UNFCCC process provides important clues to the understanding of how transnational governance operates, found both in the performativity of the negotiation process and in its documentary outcomes. As such, the Warsaw framework provides a paradigmatic example not just of the significance of those outcomes—and, therefore, of why it was considered a success—but, more importantly, of how this type of agreement enables—and is in fact a necessary condition for—transnational governance. By agreeing to *keep talking*, parties are kept enrolled in the negotiation process and in the narrative that authorizes the transnational governance of the climate. The continuous enrollment of parties not only authorizes this form of transnational governance but also allows for its gradual expansion, as other agreements are reached on different matters related to climate change. Simultaneously, while agreeing to keep talking, parties have also gradually enabled the subtle and implicit shift of the terms of the conversation, transferring the responsibilities for global climate change from industrialized nations to the Global South.

While just *agreeing to keep talking*, the UNFCCC has been able to assert its legitimacy to be the transnational governing body of the global climate, defining the appropriate approaches to solve climate change, while also defining the terms to be used when talking about it. Thus, the UNFCCC sets the language and stabilizes it, promoting the continuous enrollment of parties and precluding alternative terms that could change the overarching narrative. By agreeing to keep talking, the UNFCCC has also been able to maintain a negotiating architecture that is based on previously established power hierarchies hardly dismantled. This architecture underwrites a narrative about climate change that exempts developed countries from their responsibilities in the problem of deforestation. But more importantly, the Warsaw Framework for REDD+ and its agreement to keep talking endows REDD with potentiality—the potential to become real and to take the shape that each party can imagine. Here lies the real success of the agreement: because of its vagueness, it enables REDD to be real in a way that corresponds to different parties' expectations. That is why REDD acquires so many different forms in Acre and in Mozambique, as I will show in chapters 4 and 5. But before that, it is necessary to understand the history of how REDD came to be imagined inside the UNFCCC—the subject of the next chapter.

Chapter 3
Forests at the Center of Climate Change Policies

In 2005, the same year the Kyoto Protocol entered into force, Papua New Guinea and Costa Rica suggested a mechanism for reducing emissions from deforestation during the 11th Conference of the Parties, held in Montreal. The proposal was introduced on behalf of the newly formed Coalition for Rainforest Nations (CfRN)[1] and called for the inclusion of emissions from poor countries' forest sector in a post-Kyoto agreement to start in 2012 (Long, Roberts, and Dehm 2010). According to these proponents, deforestation and forest degradation greatly contribute to the overall GHG emissions (18 to 20 percent in some estimates);[2] therefore, reducing emissions from tropical forests alone could offer the potential to mitigate a major source of global emissions at relatively low estimated costs. This was not the first time that forest conservation was brought into the UNFCCC discussions as a mitigation strategy, since it had been suggested by the Japanese and American delegations many times in the past and pushed for by the international NGO The Nature Conservancy since the first COP, in 1995, "as a critical means of immediately dampening the rise in greenhouse gases" (Zwick 2018).[3]

At that time, the several proposals for avoided deforestation were consecutively rejected due to several scientific unknowns, and even the Kyoto Protocol only allowed for very limited participation of the forest sector. Forests were part of the Protocol exclusively through reforestation and plantation activities (called afforestation) within the clean development mechanism.[4]

However, 2005 was the first time that parties agreed to bracket on a series of scientific unknowns in order to incorporate an instrument to prevent deforestation based upon a financial compensation. By attributing the "right price" to forests, REDD would be successful in stopping deforestation and break a long historical trend of failed policy efforts, they argued.

In this chapter, I trace the history of REDD within the UNFCCC negotiations, to highlight how forests came to be at the center of climate change policies and how that process brought to the fore some significant paradoxes. Among these, the constant need to uniformize and quantify what is inherently diverse and unquantifiable, to the point of rendering antienvironmental a mechanism that purports to be about environmental conservation, or the contradictions between market advocates' ambitions regarding REDD, and markets' own rules. Despite these paradoxes, REDD constitutes a powerful example of the successful articulation between the neoliberal imperatives of transnational governance with national policymaking.

The concept of neoliberalism can be problematic due to the existence of multiple and, at times, contradictory definitions of its meaning (Brenner, Peck, and Theodore 2010; Flew 2014), which requires the examination of neoliberalism always in its specific contexts. Notwithstanding, understood as both a structural force and an ideology (Ganti 2014), neoliberalism can be generally characterized for prescribing an increased privatization and marketization of public services, the commodification of elements that were once accessible (what is usually referred to as "the commons," such as a forest), and the decentralization and handover of governance to non-state actors such as NGOs and international organizations (Castree 2008; Fletcher 2010; Fletcher and Büscher 2017). Despite these general features, neoliberalism must be conceived as a dynamic concept, always in the process of adapting to different circumstances, incorporating new elements and characteristics.

Drawing on other authors who have examined how neoliberalism operates in the context of environmental conservation, I contend that the introduction of REDD in the UNFCCC negotiations expresses the growing neoliberalization of the policy approaches to the problem of climate change. As it was introduced in the negotiations, REDD clearly embodies an inclination toward the privatization, commoditization, and financialization of the forests of the Global South, the choice of markets over regulation to curtail deforestation, and the predominance of international environmental NGOs in the implementation of policies to manage the forests. In this context, the increasing prevalence of environmental NGOs—especially large international NGOs—in environmental governance corresponds to "the creation of new spaces of government in which local, national, and supranational actors can operate, but which are outside or beyond the sphere of democratic politics and public accountability" (Shore 2011, 127). The emergence of these "new spaces of government" also constitutes a feature of neoliberalism, rendering the policies emanated by them harder to challenge given their operation outside of the sphere of democratic accountability.

Although REDD corresponds to the subscription of a neoliberal ideology, that does not mean that REDD has been received, understood, and implemented in coherent or even similar ways across different locations. Indeed, as the REDD template began to circulate from the UNFCCC platform into national and local levels, it generated a potent and varied imaginary that has ultimately underwritten the creation of multiple policies, some of which are at times inconsistent with and even contradictory to neoliberal goals. As it continues to be debated, the REDD template is far from being monolithically defined or even understandable simply as a forest conservation mechanism (Hein 2019, 4). At the national and local levels, it is constantly being redefined, reinterpreted, and reconfigured—while at the international level, it is represented as always operational and coherent. Usually, REDD is simply defined as a mechanism to financially compensate avoided deforestation, but as my ethnographic encounters reveal, that does not tell the whole story.

Including Forests in the Negotiations

The inclusion of forests and other land use activities in the Kyoto Protocol was never a consensual matter and raised many questions regarding how to count existing carbon sink capacities through reforestation or afforestation. Even though most agree on the environmental benefits of planting trees, there are still doubts about how to ensure that such an endeavor really reduces the amount of GHG in the atmosphere, exactly how much, and with which degree of permanence—all without creating negative impacts on the environment or local communities. Given all these uncertainties, key terms such as "forests" or "afforestation" were not even defined in the Kyoto Protocol (Fogel 2005, 193). In order to include forest activities in the emissions reduction effort, it was thus necessary to agree on what a forest is, and what kinds of human interventions in forested areas might be beneficial or detrimental to both the environment and forest-dependent communities.

Only in 2001, during the Marrakesh Accords, were these terms agreed upon, ensuring that the inclusion of forest activities would "not undermine the environmental integrity of the Protocol" (Yamin and Depledge 2004, 123). These accords were, therefore, significant because they introduced a definition of forest and provided some clarity on the methodologies for the inclusion of land use activities in emissions reductions.

Despite the agreement achieved in Marrakesh, many other key issues regarding forest carbon remained unresolved (Yamin and Depledge 2004, 124), namely those involving the accurate measurement of sinks (how much carbon they can retain), the permanence of those sinks (for how long),[5] and the possible increase in deforestation in areas adjacent to those being protected.[6] Equally unaddressed were the risks regarding possible impacts on biodiversity and on forest communities when, for instance, reforestation endangers food security by taking place in areas used for agriculture. The rules for afforestation and reforestation projects under CDM were defined after the 2003 COP, but forest projects remained almost negligible in the overall CDM pipeline.[7] This was due to persistent risks inherent to forest conservation and the impermanence of the emission reductions generated through these kinds of projects.[8]

Scholar Cathleen Fogel (2005) describes how in one IPCC meeting on land use and forestry (prior to the emergence of the REDD proposal), all the scientific and technical uncertainties of the forest sector expressed by African and European authorities were continuously dismissed by the United States, Japan, and Canada delegations, who strongly pushed for the inclusion of such projects in the Kyoto Protocol. Forest carbon sequestration projects were controversial from the beginning because they were untested, poorly understood, and fraught with scientific uncertainties. Moreover, "some scenarios appeared to allow industrialized countries to claim nearly all of their emission reduction commitments through ... ongoing, 'business as usual' forest management activities" (Fogel 2005, 193). This possibility concerned European negotiators for its potential to undermine efforts to reduce emissions originated by fossil fuels, since the forest sector presented itself as a cheap mitigation option.

The group of African countries pushed for the inclusion of considerations on the social, economic, and cultural impacts of such activities, and "despite a large existing literature on social problems associated with tree planting projects in developing countries," African countries' claims remained unaddressed (Fogel 2005, 203). For African authorities, forest-related projects were a painful reminder of similar colonial enterprises, characterized by the exploitation of resources and labor (Sodikoff 2012), and violent encroachment and land-grabbing (Hughes 2006; Moore 2005) for the benefit and welfare of Western nations.

When REDD was formally introduced as a mechanism to reduce emissions, the same uncertainties regarding carbon measurements and the permanence of sinks continued to raise many questions. Unlike afforestation and reforestation projects, REDD was about *preventing deforestation* (i.e., reducing emissions through an avoided action), which created additional interrogations over how to account for such an abstraction. The idea behind it was quite simple: a forest baseline would be established, as well as a scenario of current and predicted deforestation. The financial compensation would reward the difference between the deforestation predicted and the deforestation actually avoided. Although all reducing emissions projects require some sort of temporal abstraction between a present situation and a predicted future, REDD increased the level of this abstraction. Indeed, what would be financially rewarded was not the implementation of a certain action (i.e., a project), *but the avoidance of an action* (i.e., deforestation). Moreover, REDD relied on contested methods of carbon accounting, and, as in other projects, enabled the construction of fictitious baselines and scenarios. For these reasons and sovereignty concerns, Brazilian authorities were adamantly against the inclusion of avoided deforestation in the CDM and were able to prevent it until 2005 (Fogel 2005, 205). Finally, REDD was proposed for implementation in developing countries which, according to the Kyoto Protocol, should not be responsible for efforts to reduce emissions. This reflected a change in the spirit of Kyoto of *common but different responsibilities*. The reception of REDD by some of these countries was, therefore, not enthusiastic even though the proposal had been introduced on behalf of two developing countries: Costa Rica and Papua New Guinea.

Behind these countries' proposal, however, was the head of the CfRN, Kevin Conrad, a business academic from Columbia University, who "had an interest in securing revenue for Papua New Guinea from carbon credits" (Long, Roberts, and Dehm 2010, 227) and in expanding the role of financial markets in the environmental sector. The CfRN proposal was that once a national baseline determined current and projected future rates of deforestation, developing countries would be rewarded for the preservation of their forests, and the amount of the reward (i.e., the value of their forests) would be defined by markets. This market logic appealed to officials from developed countries, who saw REDD as an affordable way to comply with their emissions reductions' commitments, instead of implementing more expensive options to reduce both their levels of industrial pollution and consumption of fossil fuels. By this time, European authorities were no longer opposing the introduction of avoided deforestation as a mitigation mechanism—so long as it would not replace all other efforts to reduce emissions:

> The EU "considers that REDD+ verified emissions reductions could be used in the medium term for compliance *subject to strict quantitative limitations* and in the medium to long term be phased into the international carbon market." (Parker et al. 2009, 44)[9]

During the Bali COP, in 2007, REDD gained traction within the negotiations with the establishment of a specific REDD working group and the inclusion of tropical deforestation in the Bali Action Plan. This plan was not clear on whether REDD would be a market-based mechanism or if it would follow the same logic of conditional aid. However, among developing countries, there was an understanding that compensation for the protection of forests in countries of the South had to come from "countries of the North (not carbon markets), and that the carbon saved from avoided deforestation projects could not be used as an 'offset'" for industrialized countries (Long, Roberts, and Dehm 2010, 228). In other words, REDD was acceptable for poor countries if conceived as an aid mechanism, rewarding the conservation of forests, and could not be a license for industrialized countries to continue polluting. All the while, the scientific uncertainties and potential impacts related to the implementation of REDD remained undefined.

The Politics of Scientific Uncertainties

The question of whether REDD should be an exclusive market-based mechanism or a fund to reward successfully avoided deforestation became the biggest obstacle in the negotiations (Dooley 2010, 2), leading to a deadlock during the 2009 Copenhagen COP. However, and because the whole conference was marked by polemic events generating highly ambivalent results in what regards providing solutions to the problem of climate change, "REDD received more mentions than any other climate mitigation strategy in the controversial Copenhagen Accord" (Long, Roberts, and Dehm 2010, 229). This greatly increased the enthusiasm of REDD advocates over its future adoption as a mitigation strategy. In 2010, in a cautionary report about REDD, the environmental NGO Friends of the Earth noticed how REDD was becoming increasingly important outside of the UNFCCC floor:

> Although not yet agreed, REDD is already generating considerable momentum in the "real" world outside the halls of negotiation. In fact, many negotiators probably consider it one of the more successful aspects of the UNFCCC talks (which are generally characterised by political footdragging and intransigence). (Friends of the Earth International 2010, 5)

Despite the momentum, issues like leakage; how to measure, report, and verify (MRV) forest carbon; and the market versus nonmarket approach continued to stall the REDD negotiations. The Doha COP, in 2012, was particularly tense with respect to MRV, with developed countries, led by Norway, insisting on additional procedures for reporting and verification of emission reductions from REDD before transferring funds—or what my Brazilian interlocutor at the Environment Ministry called *the MRV of the MRV*. Developing countries understood this demand as evidence of the lack of trust that threatened to undermine the negotiations. Brazilians in particular accused Norway of deploying a tactic to delay the disbursement of funds and push for the establishment of REDD as a market mechanism—something that poor countries had always opposed. Some small environmental NGOs, like the Forests and the European Union Resource Network (FERN), considered developed countries' demands unfair given the urgent need to cut emissions in those countries, the inherent inaccuracy of forest carbon verification, and the complexities, uncertainties, and high costs associated with verifying emission reductions from forests (FERN 2012; Dooley 2014).

In the beginning of 2013, some of the observers I talked to about these negotiations confided their disbelief in the possibility of solving the issues that were blocking the REDD progress, namely finance and the MRV process. Many feared that REDD would lose its momentum. A person working at that time in the forest program of the environmental NGO WWF, based in D.C., manifested his frustration this way:

> So, in the policy side of things, I think we lost some of that momentum too, I think REDD runs the risk of alienating itself from the rest of the mitigation discussions. That it was doing really well as a front-runner and it needed to be that, to get where it is now, but if it continues to separate itself from other mitigation, finance. It actually won't maintain that momentum, because people can't push these things along separate tracks. I think it's just very politically challenging to think about how you finance REDD, versus how you finance the rest of mitigation. Why is it different? Why do we need to set different rules for monitoring, reporting and verification? They should be the same set of rules that we apply to all mitigation in developing countries. There was a political momentum behind Copenhagen, that we've never seen before, and we may never see again.[10]

If, for this practitioner, the loss of REDD momentum could mean the impossibility of seeing the mechanism inscribed in the climate change policy toolkit, for those involved in the voluntary carbon markets—already developing some types of REDD projects—the delayed negotiations meant lower prices for forest-related offsets and increased risks for this kind of investment. Hence, the disappointment expressed by a member of a company responsible for the development of REDD projects for the voluntary market:

> I just don't see a globally binding UNFCCC new treaty, replacing the Kyoto Protocol with REDD. I mean, maybe, ideally that would be the best.
> *Question—So 2020 is not going to happen?*[11]
> Maybe, but that is too far off for REDD to survive. I would say. A REDD project created today, cannot wait till 2020.[12]

Despite the low expectations and the fears over REDD losing its momentum, the Warsaw COP, held in 2013, came to be known as the REDD COP by putting forward the Warsaw Framework for REDD+ (see chapter 2), a document that due to its capacity to "enroll" parties, was able to transcend the main negotiating obstacles and, therefore, was deemed a success (Mosse 2005). This agreement was not, however, the product of any scientific revolution bringing more clarity to the issues that are still hovering above forest carbon accounting, nor more accurate information about the possible socioeconomic and environmental impacts caused by REDD projects. Measuring forest carbon remains an uncertain and far from accurate task (Dooley 2014; Disney et al. 2018). Rather than a change in the scientific paradigm of forest carbon accounting, what changed was the political context, which enabled parties to reach a political agreement on how to address these issues—even if (or perhaps because) the science remained ambiguous.

Scholars Farhana Yamin and Joanna Depledge refer to many of the questions regarding forest carbon accounting as "policy issues," while also recognizing that this is an "otherwise highly complex area where scientific and technical issues can all too easily obscure the fundamental policy concerns at stake" (2004, 124). Scholars of science, however, explain that many public policy problems "are complex 'hybrids' of the scientific and the political" with destabilizing effects (Hilgartner 2000, 4), and that "scientific uncertainty and the pressures of decision making [often] lead to a forced marriage between science and politics" (Jasanoff 1990, 8). In sum, solutions to the problem of knowledge are frequently (if not always) political, and "the knowledge thus produced and authenticated becomes an element in political action in the wider polity" (Shapin and Schaffer 1985, 342).

As such, the little knowledge available concerning forest carbon accounting was nonetheless deployed by political leaders from developed countries to support REDD as part of the climate change policy kit, against the doubts of poor countries' delegations. More specifically, these political leaders adapted knowledges and techniques from the timber industry and asserted them as settled science in carbon accounting—in a political process. A paradigmatic example of this process was the decision to use the allometric equation to measure carbon. This equation has been used by the timber industry to measure the volume and biomass of forests, and it entails the measurement of trees' diameters within a specific forest sample. The choice of using the allometric equation to measure carbon can only be explained by the fact that those championing REDD needed to have a workable accounting method to operationalize the mechanism (Lovell and MacKenzie 2015, 87). The problem is that while the equation is relatively accurate to estimate timber volume and biomass for single-species forests, trees of about the same age, and trees from a single geographic area, when converted to measure carbon in biodiverse, multi-aged forests in locations from which there are no samples to compare with the original equation, the results present significant margins of error (87). Associating the allometric equation with more recent techniques of remote sensing reduces these margins of error, but measuring forest carbon is still far from being an accurate activity (Dooley 2014) and keeps puzzling scientists (Vieilledent et al. 2018; Riley, Zhu, and Tang 2018).[13] There is also the permanence of sinks, which is unpredictable, given the possibility of fires, diseases, or unforeseeable deforestation.

Notwithstanding these problems, in a political context in which deforestation is generally perceived as bad (especially in tropical forests, considered to have greater capacities to retain carbon),[14] in which increasing deforestation is tied to climate change, and in which developing countries have been gradually called upon to take the responsibility for climate change, the gaps in knowledge about forest carbon accounting were not significant enough to keep REDD at bay. Where science gaps existed, these were resolved in ways that supported the objectives of industrialized countries, based on a political reasoning. The political decisions taken upon the controversies around REDD informed the terms of the debate to the extent that precluded alternative views over knowledge gaps. Finally, once closed, these gaps have been treated as definitely resolved. They are no longer open for debate.

My point here is not to decry that all these scientific uncertainties about forest carbon were swept under the carpet of negotiations through political decisions. What I intend to highlight is that none of the issues that prevented forestry from having a strong position in the CDM back in the 1990s, when the Kyoto Protocol was being negotiated, were resolved in 2005. However, by that time, there was a different political context that enabled the REDD proposal to be accepted, even by those who were adamantly against it, like Brazil. Even though controversies and scientific uncertainties remained, the efficacy of REDD as a mechanism that reduces emissions stopped being questioned altogether within the negotiations. My conversation with a specialist in carbon accounting working for EDF, a large US-based environmental NGO illustrates this situation:

> *In the Marrakech Accords, deforestation and forest degradation projects were excluded in developing countries. So how did we arrive at REDD? What changed? Because it wasn't that long ago, right?*

Right, right. I think there was a widespread recognition that deforestation was playing a big role in emissions globally, and that we ought to do something about it. But because of the common but differentiated responsibilities (CBDR) clause in the UNFCCC convention, you know, there was no expectation that developing countries were to do anything to contribute to solving the climate change problem. And REDD was sort of the first issue to come on and sort of change that perspective, the way it was seen, you know. We can't really solve the climate problem unless we tackle this deforestation emissions. So, then people were thinking about what can we offer developing countries in order to get them to take actions without violating this differentiation between developed and developing? And, things have been progressing along the way, so, some of the developing countries are looking a lot more developed these days, and some of the, you know, what were economies in transition in the nineties, you know, didn't, they still haven't transitioned and become, they're still not doing well economically, let's just say. So I think it was around 2005, and Steve Schwartzman, who is at EDF, played a part in this, the Papua New Guinea negotiator Kevin Conrad played a part in this, they actually came forward and said, *we would like to have a mechanism that allows for compensation for reductions*, so it was called compensation reductions at that time—where we would pay developing countries on a sort of per time basis, for every time that they reduced, and that would offer a financial incentive for them to take action, to be financed by developed countries, so the CBDR was still preserved—at least that was the thinking.[15]

By 2005, when REDD was brought to the negotiations, industrialized countries were no longer willing to acknowledge their higher (if not exclusive) responsibilities for the problem of climate change, nor were they willing to bear the brunt of the mitigation costs alone. At the same time, the negotiations grew responsive to the rising tide of neoliberal beliefs that came to characterize so much post-Cold War policy thinking, particularly, the faith that markets could provide affordable and effective solutions to many problems, including the problem of climate change (Harvey 2005). Within this global geopolitical moment, REDD emerged as the perfect tool to both involve developing countries in mitigation efforts and demonstrate the efficacy of markets in the resolution of environmental problems—despite all the knowledge gaps about forest carbon.

Roadblocks in the Negotiation Trail

MARKET OR NONMARKET?

The discussion of whether REDD should be an exclusive market mechanism or a rewarding system for avoided deforestation has been one of the main contentious issues in the negotiations. While countries like the United States, Japan, and Canada pushed for the former, Brazil and other developing countries preferred the latter. There are several important distinctions between the market and the system of payments per results that was being proposed. While the market implied that the trade of the forest offsets in a platform opened to all kinds of buyers (governments, corporations, private investors, etc.), payments per results would consist of operations between governments—either bilaterally or multilaterally. Furthermore, while the prices of those offsets would be determined by the laws of supply and demand—thus providing no guarantee on whether the prices could cover the cost of keeping forests standing—under a rewards system, the amounts transferred in exchange for avoided deforestation would be previously negotiated between the parties involved, which meant that these amounts could be subject to additional conditionalities. Finally, some developing countries also claimed that, even if they were to accept REDD as a trading mechanism, there were costs related to its implementation (like forest monitoring systems) that would have to be supported by developed countries.

Closely related to this question was the issue of private versus public financing, with developed countries arguing for the involvement of private entities in the fight against climate change. For developing countries, however, the possible involvement of the private sector was not only politically unacceptable in the sense that it represented an avoidance of industrialized countries' responsibilities, but it also emptied REDD of what they considered to be its most important element: the enforceability of state-backed agreements within the legitimacy of the UNFCCC legal framework.

Despite all the enthusiasm of countries like the United States, Japan, or Norway over the possibility of developing an international market based on forest carbon, rendering such a market real, turned out to be more complicated than imagining one. For REDD to work, the price of forest offsets would have to be higher than the revenues obtained from any other uses of forests, like timber or expansion of agriculture (Fletcher et al. 2016). That is, trees had to be more valuable standing than chopped down. However, the persistently low carbon prices registered in the markets anticipated equally low prices for REDD offsets, evincing the potential difficulties of a REDD market. Moreover, the amounts pledged by developed countries to support the creation of a REDD market also fell short of what had been deemed necessary to sustain—at least initially—the implementation of structures like monitoring systems or trading authorities. The impossibility of making REDD a market became apparent.

The solution to these obstacles was found in implementing several REDD experiments across different countries under bilateral agreements (as is the case with the agreements Norway has celebrated with several Latin American and African countries) or within pilot programs, such as those created by the World Bank and the United Nations. For REDD advocates, the multiplication of different REDD experiments would not only popularize the mechanism in different regions, but also help to set the necessary conditions for a future forest carbon market through a process of "learning by doing." This logic provided the opportunity for the private sector and the voluntary markets to expand their REDD projects under the rationale that these projects would contribute to that learning process through the accumulation of experience.

However, for developing countries it was important that such REDD pilot experiences had an institutional framing—and not be solely managed by market agents. As such, in 2008 the United Nations Environment Programme (UNEP), the United Nations Development Programme (UNDP), and the Food and Agriculture Organization (FAO) created the UN-REDD, with the goal of preparing countries for REDD's implementation. Through various pilot experiences, the UN-REDD would develop a common approach, methodologies, tools, and guidelines for REDD.[16] In the same year, the World Bank created the Forest Carbon Partnership Facility, which has a readiness fund and a carbon fund to remunerate countries according to verified emissions reductions achieved.[17] Simultaneously, individual governments, notably Norway, Germany, the United Kingdom, and Sweden, began bilateral arrangements with several countries from the Global South to develop REDD pilot experiences, envisioning them as precursors for a countrywide implementation. The government of Brazil, on the other hand, established the Amazon Fund[18] to manage the funds coming from these bilateral agreements, as well as domestic funds intended to reduce deforestation. The goal of the Amazon Fund was to help the country accomplish a target of reducing Amazonia's deforestation by 80 percent below its historic baseline over a period of ten years through several public policies deemed appropriate to prevent, monitor, and combat deforestation, which can include simple money handouts to poor communities living in forested areas. In the words of an officer from the Brazilian Ministry of the Environment I interviewed,

> We do not understand REDD as a model similar to the Clean Development Mechanism. We understand that REDD+ require[s] a national strategy, or at least a regional strategy: for Amazonia, for the Atlantic Forest, there are differences. For some regions it will be more important to increase the [carbon] stocks, in others it will be reducing deforestation—but we understand it as a set of policies, programs, projects, something more general and that will allow us, from a verified result, to obtain international financial rewards. That is our reading.[19]

WHERE WILL THE MONEY COME FROM?

Following this bilateral trend, the Durban COP, in 2011, established that "finance for results-based actions could come from a wide variety of sources, including public, private, bilateral, multilateral and alternative sources" (Global Witness 2012, 17). This meant that even though developing countries were successful in keeping a REDD market out of the negotiations, they were, nonetheless, unable to maintain REDD under the exclusive UNFCCC's institutional framework—REDD was now being tried by private entities and environmental NGOs. Billions of dollars for REDD have been pledged since 2008, but no one "really knows how much of that money has actually been deployed—let alone where and how" (Forest Trends 2013).[20]

This situation was always contested by Brazilian and other developing countries' authorities, who demanded predictability, stability, and additionality of funds for REDD—something that could only be achieved within the legal framework of the UN Convention. Furthermore, all funds disbursed in a bilateral logic, or outside of a transnational framework are conditioned to specific rules defined by a single agent (the donor country) that do not necessarily follow the principles agreed within the UNFCCC. In 2013, the same Brazilian officer described the situation in these terms:

If we don't create an architecture inside the Convention, we will be at the mercy of these bilateral agreements, which for Brazil—

Question: Do not provide security, right?

Exactly! To Brazil it probably wouldn't have such a big of impact in the short term because, well, the Amazon Fund works and it's out there. But that is not the question. The question is that there are no guarantees, no predictability [of the money available for countries to implement REDD].[21]

She went on to explain how most of the funds pledged and disbursed for REDD were not additional at all and were often double counted, meaning that money disbursed for a certain development goal was being accounted also as REDD money:

This is confidential but I'm going to tell you who it is. Japan, he grabs all his finance that has something to do with forests or vegetation and puts it under REDD. Japan takes all his *ODA*[22] and does *re-labelling*, and then says it's REDD. And Japan emerges as the biggest REDD donor! Imagine how Norway feels when looking at this [the REDD+ database]? This has to be a joke! Other countries do the same. Germany also takes all her funds related to biodiversity and puts it under REDD, or climate. Once we have a [financing] architecture inside the Convention, Japan will not be able to say that the *earthquake relief loan* that he gave to China is REDD! So to us, this is a scandal! But if you talk to other developing countries, they are in a position of dependence towards donors. Brazil doesn't have that. We do not submit ourselves to those thousands of requirements and for that reason we are not in UN-REDD or the FCPF.[23]

The privileged position of Brazil compared to other developing countries was made clear to me several times while I was in Mozambique; when mentioning what was going on in Brazil in order to evoke comparative comments on the Mozambican reality, I would often hear outpourings like, *Brazil has enough money to do whatever they want—unfortunately, we have to comply with donors' demands.* Indeed, when I talked to several Brazilian authorities about the Amazon Fund, all of them stated that the Amazon Fund *is* Brazil's REDD (even if the money was applied in ways that would not always conform to the usual donors' demands). As it is structured, the Fund allows authorities to manage their forests as they see fit, to the extent that Brazilian authorities can demonstrate success in achieving the contractual goals of reduced deforestation. Unlike what happens under the UN-REDD or the FCPF programs in which countries must follow given templates (in terms of policies, institutional arrangements, and activities), Brazilian authorities are free to manage the Amazon Fund according to what they think is the best course of action to reduce deforestation. Countries under other REDD programs can only follow the rules dictated by those same programs, which ultimately aim at a future international market.

Intrigued by why Norway would finance Brazil directly, allowing its government to implement REDD as a public policy, while on the UNFCCC floor the group of developed countries (including Norway) was pushing for a market-based mechanism, I questioned the representative of Norway's embassy in Maputo about his country's position on REDD finance. During our conversation, he clarified that his government[24] was not favoring a market-based approach over any other form of funding. Instead, Norway was trying several approaches and learning from successful experiences—Mozambique was one of those experiences.[25] Later, I was able to interview an officer from the Norwegian government, in Oslo, who clarified the special position of Brazil in the context of their foreign cooperation: Brazil is a "unique tropical country," she said, with a "stable democracy [and] stronger institutions" that allow it "to make political decisions and then actually be able to carry them out"—less achievable in "other weaker countries." Moreover, "Brazil has a large middle class that pushes for environmental protection, and that is not always the case in other lesser developed countries." The language here is revealing: Brazil is a tropical country, yet a civilized one; therefore, it is allowed to implement its own policies. Then, she added that for Norway, the forestry sector

> is the most prestigious and the biggest development policy initiative. It represents around 10 percent of the cooperation budget, so it is a very high-profile initiative. Norway is quite far the largest donor when it comes to REDD, and it has an important goal in the Norwegian development aid agenda.

This same officer asserted that Norway does not look at REDD as market-based, but in the negotiations, I think there are a lot of different views of what REDD should be, and maybe the market-based approach is how it is perceived by a lot of countries, and that is something that is creating a lot of discomfort for some countries that believe REDD is all about selling nature and putting a price on forests. But that is why maybe a lot of countries are against REDD as a whole, but I think the way that we approach REDD, being more of a national approach, a large reform program, has nothing to do with selling, selling trees. That's not what it is about.[26]

If for the Norwegian government the sources of finance for REDD did not have to be strictly market-based, that assertion was most likely caused by the realization that such a market would hardly become a reality—at least in the short term. Although I conducted these interviews in 2014, Norway's submission to the UNFCCC on Views on Results-Based Finance for REDD+ in March 2012 had already noted that "the lack of a credible demand [for such a market] . . . is the most critical barrier to increasing REDD+ actions and results," adding that "if we are serious about the 2 degree target, ensuring a prize on carbon and establishing a credible demand is urgent" (Norway 2012). At that time, it was pretty much agreed among the parties that nonmarket funding was necessary to prepare the groundwork for REDD's implementation (or to establish a *credible demand*), after which some form of market or payments for results would follow.

Among the international environmental NGOs that support REDD (in Washington, D.C.; Brazil; and Mozambique), the positions were not much different: financing was mostly presented as desirably mixed; that is, public finance should be applied in the creation of legal frameworks and rules, land regulation, methodologies, and infrastructures, and after that, private finance should follow, with the implementation of REDD projects whose final goal would be to generate offsets to be traded in international carbon markets. Despite the nuances, parties from the Global North and international environmental NGOs nurtured the ambition of having a global market for forest carbon offsets.

The overall agreement over the need for public funding to at least get REDD started follows from the confrontation between a neoliberal ideology that deems markets as efficient ways to address policy problems, and the concrete rules of markets themselves. Because international market flows are based on commodity chains and extractive activities that are more profitable than "conservation markets," there is no demand for REDD. The idea of providing funds to establish the needed infrastructures to implement REDD is based on the assumption that once these things are in place, the price of conserving trees would exceed the profits of cutting them down. However, such an assumption "is untenable, which is why in practice conservation markets become increasingly less market-like over time and must incorporate forms of subsidy or regulation antithetical to their original aims in order to achieve conservation" (Fletcher et al. 2016, 674–75).

The paradox among market advocates is that while they keep endeavoring to build an international market for forest carbon, they seem to ignore markets' rules. Despite being moved by neoliberal beliefs about the efficiency of markets to replace policy regulation, these market advocates continue to imagine a market that still needs all sorts of policy intervention to become real—namely, interventions to create demand and to produce scarcity.

MEASUREMENT, REPORT, AND VERIFICATION (MRV): GOVERNMENTALITY TECHNIQUES

To create a future REDD international market, parties had to agree on a set of standardized procedures to account and report carbon, as well as its trading rules. The Warsaw Framework for REDD+ (see chapter 2) established the rules for this standardization process under the acronym MRV—measurement, report, and verification. While MRV systems were praised by the UNFCCC as necessary conditions to achieve *transparency, consistency,* and *accuracy*, parties from the Global South often complained that increasing demands for MRV only consisted of maneuvers to dodge the disbursement of funds—especially as everything under negotiation (policy frameworks, institutional reforms, safeguards, implementation) was thought to also require MRV systems in an infinite recursive process.

For authorities of developed countries, MRV systems embody the "twined precepts of economic efficiency and good practice" (Strathern 2000, 2), while ideas of measuring performance and verifying the reports of such performance are directly tied to the notion of accountability—perceived as positive value. MRV as an instrument of accountability "is but one expression of a more global process of neoliberal economic and political transformation" (Shore and Wright 2000, 58) or what Foucault called neoliberal "governmentality," in which "the role of the government [is] premised on using the norms of the free market as the organizing principles" (58), "seek[ing] to bring persons, organizations and objectives into alignment" (61). In this case, the "alignment" that MRV seeks is not just in relation to the standardization rules of carbon accounting and reporting, but, more importantly, in what concerns how forests ought to be understood and *managed* under precepts of economic efficiency.

In this regard, REDD (and its MRV) does not differ from other forms of development intervention informed by a neoliberal logic that privileges market approaches over policy regulation, and the activities of private agents over governmental actors, focusing on measurable goals and accountable achievements. This logic became dominant especially in the beginning of the 2000s, with the establishment of the Millennium Development Goals (Rist [1997] 2010, 236). But measuring REDD performance and results has revealed itself to be very challenging, demonstrating how audit procedures, far from being objective and neutral, develop classificatory and normative frameworks with powerful subjectification effects (Shore and Wright 2000, 62).

More than carbon accounting, the implementation of an infrastructure enabling REDD (i.e., measuring and reporting systems, forest reference levels, safeguards, etc.) is much more difficult to quantify and have its successes measured—as required by the UNFCCC and authorities of developed countries. Such difficulties are recognized by REDD practitioners, who acknowledge the "illusion" of developing "a purely scientific or technical performance measurement system for all aspects of success" and that "REDD+ performance indicators can vary across countries, depending on national circumstances, stakeholder views and REDD+ strategy objectives" (Wertz-Kanounnikoff and McNeill 2012, 238). From which follows the conclusion that a good performance indicator would be laws enacted and put into practice:

> Aims might include areas such as transparency, participation and rights. The actions would focus on implementation to secure the aims: specific plans, systems and *laws to be prepared, passed and implemented*. . . . Performance becomes a set of conditions to be met, with the performance indicators spelled out as clearly as possible upfront, to minimise room for varying interpretations. (243)[27]

In the end, such performance indicators constitute forms of interference in the legal system of the countries/regions wishing to implement REDD, in order to access the necessary funds to initiate the process. This is an interference that aims to put in place laws on the forestry sector that "minimise room for varying interpretations." Even though all is done in the sake of "transparency," "participation," and "rights," these kinds of interferences are informed by "older and cruder power plays . . . which undermine the core principles of democratic government" (Shore and Wright 2011, 16) and states' sovereignty. Performance indicators such as these, ensured by the completion of an MRV system, are thus governmentality techniques intended to shape countries' governance and forms of subjectification. This interference occurs mostly in the forestry sector, but it is easily extended into other domains, as I will demonstrate in the following chapters.

Both processes that I followed in Mozambique and Acre were subject to informal MRV systems during the preparedness phases of their REDD processes. In Mozambique, the government started to draft a national strategy for REDD with the help of a Brazilian NGO, but once the World Bank got involved in the process, and in order to get access to the funds made available by the FCPF, the authorities were forced to drop that document, and start a whole new national strategy—this time following the World Bank's requisites. The authorities also had to prepare a document called readiness preparation proposal (R-PP) outlining how Mozambique would develop *performance measures*.[28] Both documents were drafted by outsider consultants and not by Mozambican officials but, more importantly, they reflected the interests of potential foreign investors and donors. In the case of Acre, because the state government intended to sell offsets generated by the state's forests in the voluntary market, such offsets had to be certified by a foreign entity. To access the international market, the state government obtained funds from the Germany Cooperation Agency and followed the requisites of the agency to obtain the necessary certification.[29]

Getting ready to go through MRV thus requires a detailed and lengthy process of measuring the performance of the policymaking process, reporting its product, and waiting for a third-party verification of the content of such legal products—money flows according to the sequence of such processes. Tellingly, although the authors of the CIFOR publication cited above recognize that performance measurement for REDD's first phases are likely to be different across different countries, they end their piece with a wish for uniformization:

The growing body of experience and data on performance measurements may ultimately allow the establishment of internationally agreed standards for REDD+ performance assessment. (Wertz-Kanounnikoff and McNeill 2012, 246)

This desire for standardization does not correspond only to a bureaucratic endeavor of framing reality into previously defined categories such as "forests," "sinks," "baseline," or "stocks," it also follows principles of scientific forestry forged during the eighteenth century in connection to centralized state-making initiatives during that time (Scott 1998, 14). Under such principles, forests are economic resources to be managed under centralized planning with the goal of generating revenue. REDD fits this perspective, although its implementation is no longer connected to processes of state-making. Instead, it is now part of a wider process of transnational governance-making, in which the centralizing entity is the UNFCCC. By defining rules for transparency, measuring, reporting, and verification compliance and enforcing such rules upon funding conditionalities, the UNFCCC generates its authority as an international organization, orienting states' actions and creating their social reality (Barnett and Finnemore 2004, 6). This social reality encompasses the creation of bureaucratic rules as organizing principles for forests and their role as carbon sinks, which are standardized, systematized, and depoliticized by technical definitions of what a forest is and by scientific methods of carbon accounting.

If REDD constitutes a paradigmatic example of how neoliberal imperatives have become dominant in the transnational governance of the environment—in this case, of the problem of climate change—the MRV systems (seeking to impose standardization) are one of the governmentality techniques deployed by this transnational governance. As in any other form of governmentality, standardization and MRV subjectivizes: it has "transformative effects" (Strathern 2000, 287) by changing laws and regulations, by circulating people and specific forms of knowledge, and by affecting hierarchies and power structures. Developed countries' authorities and the UNFCCC claim that MRV is a necessary step to ensure transparency, accuracy, and consistency, even if such standardized systems of performance measurement and reporting conceal more than they reveal (Sanders and West 2003) or if the standardization that is imposed ends up erasing relevant specificities.

INFINITE MEASUREMENTS AND VERIFICATIONS: ASPIRING TO STANDARDIZATION

Implementing MRV systems as defined by the UNFCCC implies a degree of aspirational standardization that, besides being problematic by itself, is also impossible to achieve. As mentioned before, "accurate estimation of carbon storage involves so many variables and proxies, (including estimates of soil carbon, levels of degradation, stocking rates and timber variety), that unacceptable margins of error remain" (Dooley 2014). Forest carbon measurement is still a science of estimates that can be easily inflated or deflated depending on who is doing the measurement and for what purposes. I had the opportunity to confirm this statement during a trip to a private REDD project in Acre. I traveled with two American certifiers (and the also American project developer) with the goal of observing their process of certification of that project. One of the things they wanted to check were some measurements given by the project developer that did not match their own estimates. Once in the project site, and because time was short and the conditions to continue traveling were precarious, one of the auditors suggested inputting his own estimates, made at his desk and without knowing the place, instead of measuring them on site. The developer did not accept this suggestion, so they had to go to the specific area in the forest. This situation demonstrates that more important than measuring is the authority of who is doing the measurement: understandably, project developers tend to overestimate the carbon sink capacity of their projects so they can earn more profits from the sale of emission allowances; auditors tend to be more conservative in their estimates.

Current techniques of forest carbon estimates are based on older methods of measuring biomass and timber (Lovell and MacKenzie 2015), which means that these are methods built on the assumption that, first, forests are "resources to be managed," and second, ideally, these are high-yield single-species forests. Under such assumptions, what makes a rainforest so special, and supposedly what is its major environmental value—its biodiversity—becomes not only secondary, but also an obstacle in achieving the standardization that MRV requires. Biodiverse forests are less legible because they are composed of diverse tree species, of different ages and sizes, entangled in a messy vegetation, all of which are part of a nature that is "dynamic and heterogeneous, formed again and again from presences that are cultural, historical, biological, geographical, political, physical, aesthetic, and social" (Raffles 2002, 7). Such an entanglement cannot be encompassed by a single forest definition as the one put forward by FAO in 1948, and which is at the basis of this endeavor toward standardization.

After World War II, FAO's main concern was a possible timber shortage, so in order to conduct a global forest inventory, it "adopted a forest definition suitable for assessing wood harvesting potential," enabling harmonized reporting (Chazdon et al. 2016, 541). Parties willing to implement REDD are encouraged to adopt FAO's definition of forest, or at least to define forests closely following FAO's guidelines. REDD's MRV systems are thus dependent on this definition, from which baselines and deforestation rates are estimated. The problem is that a definition that was created to assess biomass volumes does not take into consideration other indicators that, in the context of climate change, can be more important, like biodiversity, or nontimber forest products.

Forest definition by FAO: Land with tree crown cover (or equivalent stocking level) of more than 10% and area of more than 0.5 ha. The trees should be able to reach a minimum height of 5 meters at maturity in situ. May consist either of closed forest formations where trees of various storeys and undergrowth cover a high proportion of the ground; or open forest formations with a continuous vegetation cover in which tree crown cover exceeds 10%. Young natural stands and all plantations established for forestry purposes which have yet to reach a crown density of 10% or tree height of 5 meters are included under forest, as are areas normally forming part of the forest area which are temporarily unstocked as a result of human intervention or natural causes but which are expected to revert to forest.

Forest definition by UNFCCC: A minimum area of land of 0.05–1.0 ha with tree crown cover (or equivalent stocking level) of more than 10–30% with trees with the potential to reach a minimum height of 2–5 meters at maturity in situ. A forest may consist either of closed forest formations where trees of various storeys and undergrowth cover a high proportion of the ground or open forest. Young natural stands and all plantations which have yet to reach a crown cover of 10–30% or tree height of 2–5 meters are included under forest, as are areas normally forming part of the forest area which are temporarily unstocked as a result of human intervention such as harvesting or natural causes but which are expected to revert to forest. (Chazdon et al. 2016)

In an open letter addressed to FAO, released on September 21, the International Day of Struggle against Tree Monocultures, the World Rainforest Movement (WRM) accused that organization and the UNFCCC of ignoring "the vital contribution of forests to natural processes that provide soil, water and oxygen" and thus promoting "the establishment of many millions of hectares of industrial tree plantations, of mainly alien species, especially in the global South" (World Rainforest Movement 2016). For the WRM, the UNFCCC's adoption of this forest definition "is just another money-making opportunity for the tree plantation industry, and a major threat to communities affected by the trend of expanding 'carbon sink' tree plantations" (World Rainforest Movement 2016).[30]

Indeed, the UNFCCC definition of forest, inspired by FAO's, "does not distinguish tropical dry forests from mesic savannas, which differ in qualitative rather than structural aspects of the vegetation" (Chazdon et al. 2016, 543), nor does it account for the restoration of existing forests.[31] Such a definition does not distinguish even between "monoculture plantation, old-growth forests, logged forests, multispecies restoration plantations, and second-growth forests," all of which are critical in conserving forests and their biodiversity (547). The disjuncture between the goal of using forest conservation to fight climate change and aspirations for standardization is apparent under a definition that is only concerned with forested areas (not species) and potential yields (not biodiversity). If the definition of forest does not take into account that certain tree species have enhanced sink capacities; and if the definition does not distinguish between a forested area that has been restored with the plantation of more trees and a forested area recently harvested for high valued timber, how can such a definition be useful in informing policies to address climate change? Or perhaps the question to ask is, for *whom* is this definition useful? And with what purposes?

What may be the most disturbing in this definition is that it does not consider logging as deforestation; that is, harvesting trees, even if illegally and without any management plan, does not constitute deforestation as this "requires a change in land use from forest to non-forest" (Chazdon et al. 2016, 543). The only kind of deforestation that under the UNFCCC's definition of forest is considered "deforestation" is a clear-cut, followed by a different land use, like agriculture. The selective harvesting of trees by logging is instead designated "forest degradation." As such, the vested interests that are destroying Mozambique's miombo forests[32] are considered "degradation" (not deforestation); whereas the slash-and-burn farming of small plots of forest practiced by poor families to grow their gardens for food do constitute deforestation activities that need to be actively fought against in the context of REDD. Also under this logic, forests in Acre are opened to the harvesting of the most valuable trees for the timber market, while rural families' slash-and-burn farming practices of small, forested areas are criminalized. The fact that such harvesting occurs in areas never opened to commercial exploitation before, leaving them dilapidated, is irrelevant in REDD schema because, under the current forest definition, it is not deforestation. Many times, I questioned authorities in Mozambique and Acre about the lack of initiatives to fight illegal logging in the context of their REDD preparedness policies. The frequent answer was that the latter was not deforestation but degradation, and REDD's main goal was to reduce deforestation. This answer is a practical outcome of FAO's definition of forest.

The WRM letter mentions that FAO has responded to this type of criticism by arguing that in order to facilitate the reporting of data, it is necessary to have "a globally valid, simple and operational categorization of forests . . . [enabling] consistent comparisons over longer periods of time on global forest development and change" and therefore, FAO's role is one of "harmonization" (World Rainforest Movement 2016) or, in other words, standardization. Although aspirations toward standardization are difficult to fulfill (due to the enormous diversity of forests across the Global South)—and even work against REDD's stated goal of conserving forests—that does not mean that efforts in that direction are fruitless or unsuccessful.

More than imposing standards, what REDD's MRV systems do is provide guidelines on what matters, while setting aside the *unimportant*. Specifically, they impose a concrete form of understanding forests, according to which national policies ought to be enacted. Therefore, the implementation of these national policies—following a given standardized template—facilitate transnational interventions by bypassing what could constitute obstacles to such interventions: the biodiversity of different forests and the specific cultural, political, and legal circumstances of each implementation site. Ultimately, more than standardization, what these MRV systems seek to do is simplify the reality upon which transnational governance (under the REDD template) will intervene.

From "Do No Harm" to "Do Good"

If aspirations for standardized methods for forests' categorizations and management can be detrimental to the purpose of forest conservation, the expansion of such logic into what is called safeguards[33] is no less problematic. Since the beginning of the negotiations to include forest carbon in the Kyoto Protocol, the African group of countries was the one expressing the greatest concerns over possible negative social, economic, and cultural impacts (Fogel 2005)—concerns only reiterated with the emergence of REDD. The worries expressed by the authorities of African countries were based on previous failed development projects entailing tree plantation or forest restoration, with detrimental effects on local populations, including their forcible evictions.[34] Recognizing the possibility that the implementation of REDD projects could have deleterious consequences for the populations, parties agreed during the Cancun COP in 2010, that all REDD activities should be consistent with a defined set of safeguards, which were supposed to be equally reported and verified.[35]

In order to be approved by all parties, the set of safeguards defined in Cancun was broad enough to take into consideration national specificities and circumstances (in fact, the expression "according to national circumstances" punctuates the entire text). However, at the same time, the breadth of that definition has also led some environmental NGOs to call for "greater guidance from the international community on the use of appropriate indicators, data collection methods, and reporting frameworks" (Jagger et al. 2014, 1). Some NGOs were particularly concerned that the lack of "detail on how countries should implement and operationalize safeguard principles" (Daviet et al. 2013, 4) could lead to the subsuming of social safeguards to simple processes of "free, prior and informed consent" of local populations with no concern for other issues (Brockhaus et al. 2014, 2), thus leading to a depreciation of concerns that should be at the forefront of any REDD project.

Developed countries' authorities were concerned with providing more guidance on the implementation of a standardized framework for safeguards for other reasons altogether. Pushing for a standardized mechanism of monitoring and verification of safeguards was tantamount to imposing stricter conditions for developing countries to access funds, thus allowing for a greater delay in the disbursement of pledged money. The safeguards can also be understood as a political tool that deploys a technical, neutral, and rational language to enable interventions in the legal and political systems of poor countries (Shore 2011, 171) under the guise of a legitimate concern with preventing harm to the environment and to local populations.[36] Unsurprisingly, the language of safeguards oscillates between vagueness (enabling consideration of "national circumstances" in their application) and specificity (as efforts are made to quantify and standardize safeguards' implementation).

Over time, "safeguards have grown from minimal criteria of 'do no harm' ... to a more proactive 'do good' planning tool aimed at promoting the long-term environmental and social co-benefits of particular investments" (Daviet et al. 2013, 3). This gradual shift from mitigating risks or negative impacts to creating positive results was facilitated by a muddled language that mutated "safeguards" into "benefits." In the case of REDD, these are called "co-benefits" or "non-carbon benefits."[37] Indeed, the Cancun agreement itself had already opened the door for this mutation by the way it was drafted:

> That actions are consistent with the conservation of natural forests and biological diversity, ensuring that the actions referred to in paragraph 70 of this decision are not used for the conversion of natural forests, but are instead used *to incentivize* the protection and conservation of natural forests and their ecosystem services, and *to enhance other social and environmental benefits*. (UNFCCC 2011, 26–27)[38]

The issues arising from this language of "incentives" and "benefits" are apparent, as stated, for example, in Brazil's highly critical submission on "methodological issues related to non-carbon benefits resulting from the implementation of the activities referred to in Decision 1/CP.16" dated March 2013:

> "Benefits" should not be mistaken for "results." Any approach that would create additional layers of requirements for obtaining results-based payments may hinder the implementation of REDD+ activities. Likewise, an approach that would attribute different values to mitigation results depending on non-carbon benefits should be avoided. . . . Furthermore, several types of non-carbon benefits fall outside the scope and mandate of the UNFCCC. Discussions regarding non-carbon benefits should be fully consistent with the respective mandates of each international regime, while preserving the primacy of UNFCCC over REDD+ . . . particular national or local circumstances determine which types of non-carbon benefit arise [and therefore] such a

diversity of cases and scenarios does not favor the consideration of methodologies at the international or multilateral level for non-carbon benefits. . . . It is important to emphasize that REDD+ countries that receive payments for results . . . may choose at their discretion to invest these resources in initiatives that contribute to the enhancement of non-carbon benefits. Brazil has been already doing so through the Amazon Fund. (UNFCCC 2014c, 3)

This long excerpt illuminates the problems inherent in the mixing of safeguards and benefits, while also showing Brazil's position regarding the evolution of the negotiations. First, there is a clear critique of the potential additional conditionality over payments on the implementation of non-carbon benefits (NCBs). There are also concerns over the possible hijacking of the UNFCCC by other countries whose authorities have clearly stated that "REDD+ finance should select and prioritize REDD+ actions that generate NCBs" (UNFCCC 2014c, 39). The submission also refutes the possibility of standardizing these non-carbon benefits given countries' diversity, thus concluding with the need for each country to decide how to manage their REDD funds. That is, while pushing for REDD as a form of national policy (implemented according to each country's needs and circumstances), Brazilian authorities also seek to resist the creation of methodologies for non-carbon benefits that act as political trojan horses for transnational interventions and governmentality techniques.

Indeed, both American and European authorities have gradually linked safeguards and non-carbon benefits in the language used in their submissions to the UNFCCC as if they were the same thing. More importantly, both have made financial support for REDD dependent on the development of commoditized benefits reflecting the market logic that undergirds developed countries' vision of what REDD is supposed to be.

COMMODIFYING "DO GOOD"

In November 2014, during an event in Washington, D.C., where diverse stakeholders analyzed the trends of voluntary markets presented during the launch of an annual report, I was able to witness the prioritization of projects with co-benefits. This was an event open to business investors, carbon traders, environmentalists, and researchers, during which the NGO Forest Trends—Ecosystem Marketplace launches its annual report on the state of the forest carbon market. The report provides data on the global market for offsets generated in the forestry and land-use sectors, including transaction volumes and values, types of projects and geographical areas of implementation, features of main buyers, and trends of the market.

The 2014 report was the first one to include indicators for NCBs generated by emissions-reduction projects developed in the forestry and land-use sector. While this inclusion was widely praised by the people behind the event, it was also recognized that *translating* these benefits into the market introduces new challenges to project developers and offset buyers. That is, the authors of the report were talking about the need to quantify the benefits in order to make them tradable. One of the speakers representing a Californian company that develops these types of projects stated that "offsets need to tell a story," adding that this is "a demand from buyers [who want] offsets that tell a story that resonates with customers and employees."[39] That is, for customers in this market, the idea that these projects are reducing the amount of GHG in the atmosphere is not compelling enough; they also want these offsets to be the symbol of social and environmental deeds, inscribed in a *good story*, preferably with characters in flesh and blood (Lovell, Bulkeley, and Liverman 2009).

The importance given to these good stories provides an analytical lens into the deliberate confusion between safeguards and NCBs present in some of the country submissions to the UNFCCC. Under a market logic, risks (or negative impacts) are externalities that need to be internalized. That is, since negative impacts are something that was not accounted for following a certain activity, the way to minimize them is to incorporate them in such activity. This form of incorporation—or internalization, using the classic economics jargon—might simply be to include the price of solving or mitigating the negative impacts caused in the production costs. But such an internalization can become an advantage if those risks are commodified too. As Ulrich Beck has pointed out, the commercialization of risks does not break with the logic of capitalist development, but rather raises it to a new stage (1992, 23).

In practice, this means that in the voluntary markets, offsets from projects with *good stories* will be more expensive than those that *just* reduce emissions (and respect the safeguards); and in REDD projects supported by governments from developed countries or by development agencies, those with NCBs will be prioritized over others or more highly rewarded. The NCBs (or *good stories*) not only increase the value of the commodity being traded—carbon offsets—but also determine which countries should get the highest amount of funds for implementing REDD. The problem presented to market advocates regarding the commodification of REDD's safeguards, however, is one of quantification. How is a good story measured? Which ones should be given the highest value? How are different good stories rendered fungible? Such concern was openly admitted during the event I described above, as when the authors of the report asked for suggestions from the audience on how to quantify this new indicator in the market. The report itself attempted some quantification of these benefits, presenting the number of people employed and "trained or involved in capacity-building," hectares under some sort of environmental protection, number of species protected, and amount of dollars spent in "livelihood benefits" and "direct payments to communities" (Goldstein and Gonzalez 2014, 28–29). However, there was no information about the duration of such benefits or, more importantly, whether concerns over safeguarding the possibility of causing negative impacts with the generation of NCBs were attended to or not. This sounds redundant but is plausible; for example, when the protection of a habitat for an endangered species (a co-benefit) is done at the expense of the eviction of a local population, or when direct payments to a community (another co-benefit) create conflicts over the distribution of that money.

Ultimately, if safeguards are commoditized into NCBs—which is already happening in the voluntary markets but is also being pushed into the REDD framework by developed countries—there is an implicit risk of causing negative effects just by seeking to generate positive ones. In time, this would require safeguards of the safeguards of the safeguards, in an infinite recursive logic. This recursiveness can be paralleled to what Michel Callon has called "overflowing." Since markets are constantly overflowing (that is, creating externalities), attempts to frame those flows (meaning, internalizing them) are also constant and always incomplete (1998, 255). Therefore, making REDD a market will necessarily create externalities that, once internalized, will generate new externalities, and so forth, in a continuous negotiation process.

A New Momentum for Forests

Increasingly, forests have been identified as a crucial element in the policies to address climate change since REDD was effectively brought into the negotiation table in 2005. That move also marked a turning point in those negotiations: not only were developing countries now expected to do more (effectively dropping the mantra of *common but different responsibilities*), but policy approaches to climate change became increasingly informed by a neoliberal perspective. While this perspective was reflected on the maintenance and expansion of market-based solutions, the obligation of poor countries to do their part was openly asserted in some of the country submissions prior to the Paris COP. The US submission, for instance, stated that although recognizing the differences "across a broad continuum of all Parties," the United States "would not support a bifurcated approach [i.e., divided between developed and developing countries] to the new agreement, particularly one based on groupings that may have made sense in 1992 but that are clearly not rational or workable in the post-2020 era" (United States of America 2015). The full inclusion of the forestry sector in the mechanisms to reduce emissions would constitute the developing countries' part in the effort to reduce emissions. Since then, discourses about the importance of forests have been consistently at the forefront of all climate change policies and conversations.

In September 2014, the UN held a Summit on Forests and Climate, bringing together "a vanguard of countries and companies with input from indigenous peoples and non-governmental organizations" to discuss an action plan for the forestry sector. The *New York Declaration on Forests* that came out of the summit pledged to halve deforestation by 2020, end it by 2030, and restore 350 million hectares of degraded forest landscapes. During the event, the prime minister of Norway noted that "science confirms that addressing deforestation is critical to limit global warming to 2 degrees Celsius" and added that "we must all do our part," urging developing countries to "enforce the necessary land-use reforms to grow their economies without destroying their forests" (Climate Summit 2014). Describing the different pledges to reduce deforestation that occurred during the event, including the Rio Branco Declaration,[40] the report on this summit stated that "events of the past year have demonstrated new momentum and a growing global partnership to protect forests and enhance forest restoration" (Climate Summit 2014). It would be no surprise that forests were going to take the main stage in the Paris COP one year later.

In a report produced by the NGO Forest Trends after the Paris Agreement, the authors stated that "the global climate agreement reached in Paris marked a historic moment for forests as they are now enshrined in international climate action," adding that "there has never been a stronger call to action by as many countries to rapidly halt and reverse the trend in tropical deforestation" (Graham and Silva-Chávez 2016, 1). However, REDD is never mentioned in the final text of the agreement, while the word "forests" appears only six times on the same page of the same article, out of a document with sixteen pages and twenty-nine articles. This paradoxical lack of words for trees and forests in a document hailed as being particularly concerned with their role against the problem of climate change is explained by the authors of the report mentioned above in a curious way. According to them, Article 5 of the Paris Agreement acknowledges the critical role of the land sector, and it

> does that—not loudly, nor lost in pages of confusing jargon and complex rules, but simply affirming what the COP had essentially agreed over the years while adding an acceptable level of direction and encouragement for greater ambition in this sector. (Graham and Silva-Chávez 2016, 4)

The key word in this description is "acceptable," as the negotiations leading to the final agreement anticipated tensions over what was acceptable or not regarding REDD. Thus, while the Coalition for Rainforest Nations (CfRN) wanted REDD+ to be defined as a market mechanism in the final agreement, Brazilians were strongly against it, arguing that REDD was already established through the Warsaw framework (FERN 2015); it was acceptable to mention and value forests and their role as carbon sinks in the Paris document, but it was not acceptable to define that role as exclusively market-based.

Article 5 of the Paris Agreement is thus the only one mentioning forests, and it states that

1. Parties should take action to conserve and enhance, as appropriate, sinks and reservoirs of greenhouse gases as referred to in Article 4, paragraph 1(d), of the Convention, including forests.
2. Parties are encouraged to take action to implement and support, including through results-based payments, the existing framework . . . agreed under the Convention for: policy approaches and positive incentives for activities relating to reducing emissions from deforestation and forest degradation . . . while reaffirming the importance

of incentivizing, as appropriate, non-carbon benefits associated with such approaches. (UNFCCC 2015, 4)

These two paragraphs encapsulate and summarize the contentious issues that I have highlighted here as being the main roadblocks in the REDD negotiations—funding, verification of results, and NCBs—yet without falling on either one of the sides. As it is, REDD can be a market mechanism while also being a national policy financed as positive results are achieved. There are some rules over what to measure, report, and verify so funds can be disbursed, but the level of standardization and quantification of things to undergo MRV is proportional to the level of commodification involved in REDD activities. In short, the closer it gets to a market approach, the higher the level of quantification, measurement, and standardization. NCBs are desirable and should be incentivized but are not (yet) a condition to access funds or to implement REDD. This is how multiple and conflicting perspectives on what REDD can be are articulated in a very cohesive representation—that ultimately sustains REDD—such as the one achieved with the Paris Accord.

THE REAL REDD AND THE PRETENDERS

Despite achieving a vague but acceptable definition for the role of forests in the fight against climate change, the Paris Agreement did not satisfy the market hardliners. During COP23 held in Bonn, in November 2017, the CfRN introduced a new proposal on REDD that added to the compounding complexity of the REDD negotiations. The proposal included the creation of three trademarks for the following acronyms: "REDDPLUSX," described as a carbon credit brokerage, "RRU," the proposed carbon credits generated by protected forests under the Paris Agreement, and "REDD+," as well as the creation of a "gateway" to manage RRUs (Zwick 2017). The proposal was highly contested and eventually refused, as the acronym REDD+ is already being used by many NGOs, governments, and institutions like the World Bank and the United Nations. Yet, the way Kevin Conrad—the leader of CfRN—justified the proposal is revealing. In an interview on the Bionic Planet podcast,[41] Conrad argued that REDD+ should only be used in reference to activities included in the national carbon accounting reports under the Paris Agreement, and that the REDD+ description under the UNFCCC is that you have to have a national plan, you have to have a national monitoring system, you have to have a reference level, you have to write a report of your safeguards, and then you submit your results, and those results have to be independently reviewed by the UNFCCC itself, and then once it goes through that process, emission reductions are issued and put on the REDD+ Hub.... None of these projects [currently developed by several NGOs] have gone through that process, which means they shouldn't be calling themselves REDD+ projects. (Zwick 2017)

In the end, he added that the goal of the proposal was "to help the marketplace determine what is real REDD+ and to distinguish it from the 'pretenders.'" This concern over what is *real* and *fake* REDD follows from a clear intention to make REDD a full-fledged market, with trademarks, and accusations of *fakeness* toward nonconforming commodities. Ultimately, it was precisely this concern that also contributed to the rejection of the proposal, amid accusations against Conrad for trying to earn licensing fees (Zwick 2017). While the current state of the negotiations provides enough space for REDD to be what the different parties imagine it to be—which is accomplished by the vague language of the several agreements—the CfRN-specific proposal would narrow that space to the point of leading some parties to refuse to continue to talk. As it is, the acronym REDD enables multiple forms of transnational intervention in the forests and land sector of the Global South. At the same time, such interventions are only possible because the authorities of the countries targeted are also benefiting from them: on one hand, these authorities are receiving funds for the implementation of REDD activities (whatever these may be), and on the other (and perhaps more importantly), the vagueness of REDD also allows them to use that template to pursue their own agendas. This dynamic will become clearer in chapters 4 and 5.

The conversation over what REDD is will continue for years to come, and while organizations like Forest Trends believe that "market-based approaches will play a larger role in the future" (Graham and Silva-Chávez 2016, 8)—which is credible due to the opening to the private sector enabled by the Paris Agreement—others anticipate the coexistence of different types of REDD, in a "patchwork of different initiatives driven by distinct conceptualizations and associated objectives" (Turnhout et al. 2017, 9). The tensions around these issues will certainly continue throughout the negotiations, but while the conversation proceeds, what is being implemented across different places is not isolated from the dynamic of the negotiation process itself. The different implementations of REDD in multiple geographies provide important insights into how transnational governance operates within the spaces opened by the flexibility of what REDD can be, and of how it can be implemented. What is being implemented in Acre and in Zambézia under the category of REDD is different from what was initially introduced in the UNFCCC floor as REDD, and the two cases are certainly different from each other. Those differences, which reflect the tensions between parties in the negotiations, are also manifestations of local interpretations of what REDD can be, and for what purposes it can be used—while also expressing neoliberalism's heterogeneities.

Chapter 4

Operating Transnational Governance Locally

REDD in Acre and Mozambique

Introduction

In order to examine the multilevel entanglements of REDD, I begin this chapter describing the early experiments with REDD in the small Amazonian state of Acre, and then in Mozambique, demonstrating that in both locations, this mechanism was a transnational endeavor introduced by international actors, but enabled by local policies. Despite local and national authorities' efforts to build narratives asserting the native origin of the REDD experiment, this new form of governing forests was highly contested in Acre and Mozambique. While these two places have different institutional structures, what became clear is that in both, the early implementation of REDD followed from multiple forms of pressure applied by international actors. According to some scholars, the "intensification of pressure" is currently part of global conservation policies in their efforts toward the urgent need to save the environment, yet this pressure should be understood as a coercive force that has its origins in the mechanics of capitalism itself, and its constant intensification of consumption, production, and profits (Büscher and Fletcher 2018, 108).

Despite the various forms of international pressure to implement REDD, authorities in Acre and Mozambique adhered to it for their own reasons—

perhaps the most important one was the possibility of accessing significant funds to continue pursuing their own agendas. In both locations there was a significant process to create the institutional and legal frameworks deemed necessary for the actual implementation of a REDD market. However, the constitution of new legal and policy apparatus went well beyond the scope of a forest carbon market reaching other-than-forest areas. The REDD template, thus, enabled the intrusion of international elements in local governmental structures, which not only radically shaped these legal and political structures, but also had powerful subjectification effects (Goldman 2001; Agrawal 2005) normalizing several forms of neoliberalization in areas well beyond the forest sector.

Both of these processes of REDD implementation speak to the ways in which transnational governance operates simultaneously at the local, national, and international levels. In the end, even if REDD is considered a "failure" (for not preventing deforestation nor reducing emissions), the multiple activities being implemented under the REDD template are doing significant work on behalf of a market logic, and in the expansion of new forms of transnational governance.

Acre, the REDD Pioneer

> The State of Acre is a modern pioneer in tropical forest protection.
>
> Climate Focus 2013

When I started to look for a specific location to investigate the implementation of REDD in Brazil, several people told me that the state of Acre was *the* place to go. A very poor state often referred to by urban Brazilians as "the state located at the end of the world" or "the state that does not exist,"[1] Acre's marginality is defined by both its remote location on the border of Peru and Bolivia and by its late incorporation in the Brazilian Federation.[2] Despite its marginal position, I found that Acre was pioneering REDD in Brazil by hosting several private projects, while also developing state legislation aimed at statewide implementation of REDD, which would allow for trading carbon offsets in the international market.[3] At the time of my first visit in 2013, several private projects were still being developed, and one of them (the Purus project) had already been certified by both the Verified

Carbon Standard (VCS) and the Climate Community and Biodiversity Alliance (CCBA), thus enabling it to officially trade in carbon offsets. In parallel, Acre state authorities had also recently approved the SISA Law (System of Incentives for Environmental Services Law)[4] and were going through the steps required for international certification of the social and environmental safeguards for carbon markets. The VCS and the CCBA are two of the most important certifiers of carbon offsets in the voluntary markets.[5] Based in the United States, they are considered neutral parties to *monitor* and *verify* the integrity of the *reported* carbon data (accounting, reference levels, baselines) and the proper adherence to social and environmental safeguards. Without this kind of certification, project developers cannot sell their offsets in international markets, while for buyers, VCS and CCBA certifications guarantee the quality of the offsets being traded.

Through this SISA law, state authorities were in essence poised to commit the entire state of Acre to the goal of reducing deforestation—a process termed "jurisdictional REDD"—that is, REDD that covers the entire area of a significant geopolitical unit (such as a state in Brazil).[6] If, for some environmentalists, REDD's expansion from a project-scale into larger territories is positive for avoiding "leakage,"[7] for project developers, such a territorial expansion is equally beneficial for providing economies of scale.

Acre's pioneering REDD initiative was being taken seriously—both by its advocates and its detractors alike. All of my informants knew about REDD (though not necessarily all were equally clear on what exactly REDD was), and readily identified it as a policy in which the state government was invested. Overall, the local landscape proved to be extremely polarized, with those who avidly supported REDD arrayed against those who vocally opposed it. Although overgeneralizations should not be made, I found that by and large the state government, some university researchers, NGOs, and a few indigenous leaders (identified by those who opposed REDD as "the Indians who work for the government") tended to support REDD. Those opposing it included other scholars, political and indigenous activists, union leaders, the Indigenist Missionary Council (CIMI),[8] and other indigenous leaders. What struck me about this landscape was how difficult (if not impossible) it seemed to find anyone occupying any middle ground, as well as how this strong polarization had reshaped relationships—personal and professional—to the point of severing friendships and working relationships as people found themselves on opposite sides of the REDD barricade.

Another striking feature of this landscape was the very divergent "myths of origin" on each side. Each pole fostered its own narrative about the origin of REDD in Acre that, despite individual versions, exhibited a fundamentally coherent expression of that side's worldviews. There was something remarkable if almost eerie about hearing (and reading)[9] the same story over and over, but from entirely different people, following the same order of events and providing the same justifications. This script seemed most consistent among representatives from the state government when they were describing the approval process of the law enabling REDD. Yet the opposite narrative, even if less coherent in its various versions, shared common elements among different interlocutors.

The tensions between these narratives about REDD's legitimacy underscore a range of social anxieties about ownership, participation, and power in policymaking and implementation, while also speaking to the ways in which local actors have sought to appropriate REDD in the pursuit of their own goals. Through an analysis of REDD's myths of origin, I seek to tackle the "slippery textualities" of this mechanism by providing it with a highly nuanced form of local and historical indexicality (Dent 2016, 432–33) that ultimately allows those myths to point to—and signify—what at times are diametrically opposed values. Understanding how these narratives accommodate a whole range of interests and goals—that are often very different from those of the international purveyors of REDD—is key to understand why and how locals participate in international initiatives such as REDD. In parallel, the counter-narratives also reveal the interests at play and the power configurations entailed in REDD's implementation.

REDD's Myths of Origin

> Here is like a sect: those against it, say that REDD is the worst thing in the world, that you're selling your forest, that you will not be allowed to hunt, to fish, to build your house. And those for it, say the opposite, but they exaggerate by saying that REDD will solve all the financial problems that a community might have.
>
> José (indigenous leader), July 6, 2013

OFFICIAL MYTH OF ORIGIN

State government officials told me that the idea for REDD made its first appearance in the state of Acre in 2008, in the aftermath of broader international discussions about "payments for ecosystem services" and forest conservation.[10] Given that Acre is a very poor forest-state, REDD seemed to present an opportunity to access international funds (much needed in the efforts to reduce poverty and develop the state). Moreover, the fact that REDD was about forests suggested that it could possibly integrate some of the initiatives that local authorities were already implementing (or seeking to implement) in their broader strategy to develop the state.

Following some initial discussions, international NGOs promoted the organization of workshops with the purpose of training government officials on the technical and legal aspects of REDD, and drafting several documents on the potential implementation of REDD in the state.[11] One of these documents was prepared by a Brazilian private legal consultant from outside Acre (assisted by the transnational NGO International Union for Conservation of Nature [IUCN]), and later submitted to a public consultation which, according to the accounts of government officials, included "all" sectors of Acrean "civil society." This consultation process involved around forty different institutions throughout one hundred public audiences, and within the span of less than a year, collected over three hundred recommendations, which were purportedly later analyzed by state authorities. After this consultation process, I was told, authorities realized that the people did not want a REDD project, but rather, state policies applied to the whole state, and providing "incentives" for "sustainable practices of production." Authorities were particularly emphatic on these points: first, that the policies being implemented in the state followed from a wish expressed by the people during these public consultations; second, that whatever the government decided to do, it had to be on a state scale and not at a project level; and third, that the population wanted to be compensated (through state benefits) for production practices deemed sustainable.

The language here is, therefore, significant: my government interlocutors insisted in the participation of "all" sectors of Acrean civil society, and they consistently avoided the word "payments," insisting instead that the law that was eventually approved was not instituting a market *per se*, but a wider system of "incentives in return for certain practices of production." At first glance, this difference between the words "payment" and "incentive" seems to be merely rhetorical, yet, what they were trying to emphasize was the fact that they were not creating a market. Rather, they simply intended to compensate those with "sustainable practices" that could count toward reduced deforestation rates. For example, farmers who used manual methods of enriching soils instead of slash-and-burn practices should be compensated for not deforesting their properties.

After recalling the consulting process, Márcia, at that time an officer inside the Institute of Climate Change, explained this subtle difference between payments and incentives:[12]

> When we started these discussions, we realized that people didn't want that famous payment for environmental services. What they needed was to render their economic activities more, let's say, lucrative. So, after we analyzed more than 300 recommendations, we realized that it was not necessarily a payment for an environmental service, but an incentive for an environmental service, which could be monetized or not.[13]

Using the same logic, Eduardo, a federal government officer who was deeply involved in Acre's jurisdictional program stated:

> We could have called this law a payment for environmental service. We have always called it—and there is a significant difference in the word—incentive. I am aware of the critiques to the law—now read the law and tell me exactly where we are commodifying. Now, do the opposite and see in how many points we are socially distributing a benefit.[14]

Both Márcia and Eduardo emphasized the noncommodifying feature of Acre's program, insisting instead on the idea of providing an incentive to sustainable farmers and producers. Danilo—who I first met as the leader of the local office of an international NGO, but who later took a leading position in the state government—was equally careful with language. When referring to people providing these "environmental services," he called them "incentivizers" or "invigorators," refusing the market jargon of "suppliers."[15] This careful management of language was a direct response to widespread criticism—not just in Acre, but also at the international level—that schemes such as REDD and other environmental services were indeed mechanisms that commodify nature, and in ways that secure the property rights and expand the reach of capitalism and extractive industries.[16] These critiques were, in fact, central tenets of the policy's primary opponents in Acre, known as the "No REDDs,"[17] and composed of a coalition of local, national, and international actors.

Following the incorporation of the recommendations obtained during the consultation process the SISA law was quickly approved in the state assembly, on October 22, 2010, launching the implementation process, which entails the establishment of an institutional arrangement (composed of different offices and regulators), and the drafting of specific regulations (mostly for purposes of accessing funds and obtaining the required certifications for carbon trading). The documents publicizing the law and the approval process recounted this origin myth in terms that appealed to multiple audiences—international and local: while Amazonian conservation resonates strongly with international donors, authorities also appealed to locals by emphasizing certain elements of the law that speak to a so-called local identity.

In conversations with state officials, I was frequently told about this Acrean identity, purportedly marked by a particular feistiness (as in the wars waged against Bolivians), and pride (in the face of a certain neglect by federal authorities); but, more importantly, by a clear sense of the importance of conserving forests. This sense had been instilled throughout the violent struggles led by rubber tappers against the military dictatorship's plans of development, whose greatest representative was the environmental activist Chico Mendes (see box 1). It was not surprising, therefore, to find a government publication providing a rationale for the SISA law that could be located in this history of struggles for the preservation of forests, and on a special relationship that Acreans supposedly nurture with their forests. This relationship of Acre with its forests, as well as the international recognition of Chico Mendes as an environmental icon, thus became part of state authorities' political rhetoric:

> This approach [Sustainable Development of the State of Acre] has its roots in the singular history of Acre's settlement by an indigenous and non-indigenous population who built its culture and identity based in a relative isolation and in the intimate and harmonious relationship with the forest and its natural resources.
>
> The struggle of Acrean people in the defense of the forest was worldly recognized and has resulted in the creation of extractive reserves, radically changing Amazonia's settlement model. Since then, Acre has become a world

reference for local strategies of sustainable development. (Governo do Estado Acre 2012, 9)

This long passage also signals Acre's pioneering in these kinds of policies, working as a world reference for the implementation of sustainable development strategies, all of which are strongly contested by the group that opposes REDD and other forms of payments for environmental services.

Box 1

Chico Mendes: From union leader to environmentalist

Francisco "Chico" Mendes was the leader of Xapuri's rural union and a member of the socialist Workers' Party (*Partido dos Trabalhadores*). Due to his activities against the cattle and farming expansion in Acre and in defense of rubber tappers' livelihoods, he was murdered in 1988.

Xapuri had been the center of rubber trading since the Bolivian dominion of the region, and with the decadence of the rubber economy, it also became the area most coveted by cattle ranchers and land speculators (Paula 2016, 110–11). Born in Xapuri, Chico Mendes was a rubber tapper himself but, early on, got involved in the struggle against landowners and the deforestation they promoted. Mendes skillfully joined rubber tappers' demands to keep the forest standing with environmentalists' concerns about preserving Amazonia (Keck 1995). The support of international environmental movements, and later the international commotion caused by his assassination, helped pressure Brazil's central authorities to halt plans to transform the Amazon into industrial farming and cattle pastures and to accept rubber tappers' proposal to create extractive reserves. Thanks to Mendes's alliance with the American environmental movement, he was recast as an environmentalist.

The first extractive reserve was officially created in 1990, shortly after Chico Mendes was killed. Currently, there are ninety-six extractive reserves all over Brazil, totaling an area of almost sixteen million hectares. Drawing upon the already established model of indigenous reserves, rubber tappers demanded the allocation of lands that would remain under the property of the state but that they could use, thus securing their extractive livelihoods. Conceived as such, the reserves effectively prevented the division of the forest into small plots and their subsequent privatization and deforestation (Almeida 2002, 193).

Figure 6 Road to Xapuri, once the capital of rubber production; Chico Mendes is part of the city's identity. Photo taken by the author. Xapuri, 2015.

COUNTER-MYTH OF ORIGIN

Although not formally constituted as a group, local opponents to REDD are connected by social media; attend the same events; and participate in local, national, and international initiatives against REDD and other forms of what they call "green capitalism." Their reasons for opposing REDD are not exactly the same as one another, nor do they express their opposition to this scheme in the same way, given their different personal and professional stakes in this form of activism. However, their counter-narrative to the official myth of origin is remarkably consistent, presenting a detailed counterclaim to each element composing state authorities' narrative.

In their counter-narrative, they portray REDD as a foreign import and, in fact, little more than a follow-up to previous and older intervention schemes in the Amazon, such as the PPG7[18] led by the U.S. Agency for International Development and the World Bank and carried out by international NGOs like the EDF,[19] the WWF,[20] and the IUCN.[21] These are the same NGOs now involved in Acre's REDD program. The origin of REDD, according to this opposing group, can be traced back to the beginning of the 1990s, to discourses about sustainable development and the creation of the first extractive reserves and areas of natural conservation. The creation of conservation areas occurred as part of a broader process of territorial zoning,[22] whose primary aim was the implementation of forest-management plans, or what is now called sustainable forest management. In this narrative, transnational environmental NGOs coopted local and smaller NGOs through funding, effectively supporting the introduction of sustainable logging into conservation areas—all under the label of "community forest management." For this group, however, these community projects have been far from communitarian and, ultimately, disastrous given the level of environmental degradation they caused. REDD is thus a mere sequel to these sustainable forest management initiatives, carried out with the same rationale and by the same actors.

Furthermore, interlocutors in this group contested claims by the government that the SISA law involved any real public consultations, arguing that only those people who officials trusted were ever invited to participate in the sessions. They argued that the law was approved without any genuine public discussion, and that many representatives voted without ever understanding exactly what they were ratifying. Some claimed that the original draft of the law that was approved in the state assembly was, in fact, a document written in English—offering this as evidence that the SISA law was largely imposed on Acre by external actors. Details in these accounts include the key role in authorship by an American scientist, a resident in Acre for more than thirty years who, purportedly, was a CIA agent.[23] Many believed Acre was intended to serve as a showcase for international carbon markets and believed that its linkage with the state of California (one of the largest American economies) was intended to definitively establish those markets in the Southern hemisphere. Thinking in even broader terms, these opponents claimed that the whole SISA process had to be understood within the wider context of increased global commoditization of nature itself—a process they referred to as green capitalism.

CONFRONTING MYTHS

As with other myths, the authorship of these two narratives is social, which means that no one can be identified as the author. The myths have developed and been reproduced within these two groups at a more or less unconscious level of selection and interrelation of constituting fragments of these groups' experiences (Boon 1982, 241). The structure of these two myths is the same—they each touch on time of origin, authorship, events of the process (i.e., consultation, results, and drafting and approving the law), generation of value—yet they maintain a dialectical relationship of opposition and mutual constitution with each other; each element has a deictic part. Lévi-Strauss states that "each myth taken separately exists as the limited application of a pattern [or structure], which is gradually revealed by the relations of reciprocal intelligibility discerned between several myths" (1964, 13). More than stating that the elements of each narrative constitute pairs of symmetrical opposites that are deployed in mutual accusations between these two groups (either to provide legitimacy to the SISA law or to contest it), what I contend is that each narrative provides a space for policy argumentation in which REDD is reinterpreted in diametrically opposing ways.

While the first narrative situates SISA's origin around 2008, with the first discussions on how to bring REDD to the state, the opposite version locates this origin in a broader time frame that started in 1992. More than a mere temporal disagreement, then, those against REDD are deploying claims of temporal origin to place REDD within a broader genealogy of intervention; that is, far from new, REDD is but a new formulation of older intervention schemes. While the authorities' myth claims Acre's authorship of the SISA law, praising the participatory character of the discussions to create its legal framework, the opposing group accuses state authorities of continuing to submit to foreign interests to the point of simply translating into Portuguese a law that was drafted in a distant place (both spatially and culturally). Furthermore, they refuse the idea of a participatory process, instead arguing that the public consultations were merely a show, aimed at shoring up the legitimacy of a process that local people could not possibly support (if they had *really* been consulted).

Also, like other myths, these two narratives have phatic and conative functions—that is, they establish and prolong contact and appeal or express an imperative—as they help manage the bonds of sociality among these small groups, either supporting or contesting the SISA law, while also engaging with an audience of potential supporters—and future tellers of the corresponding myths (Jakobson 1960). However, if the concern of authorities with language regarding payments versus incentives, or the participation of civil society, can point to a metalinguistic function of their myth, the oppositional narrative can be seen as having a greater referential function, given its description of a specific historical context in which the SISA law emerged as a reenactment of older forms of intervention. In other words, while the official myth mostly serves a metalinguistic function of providing legitimacy to REDD, the counter-myth undermines such legitimacy by referring to a past of similar international interventions, of which REDD is just the latest iteration. Finally, both myths of origin fulfill a poetic function conveying a message intended to enroll supporters to uphold or condemn the law and reinforce each group's beliefs and values.[24]

Ultimately, these myths demonstrate that when international templates are transposed into local contexts, these templates are still contingent upon local circumstances. That is, the political legitimacy of making policies enabling REDD is still dependent on the creation of local narratives that express the interests of local stakeholders. The different claims expressed in these myths about the same history provide an entry point into the ways in which a variety of local actors engage in REDD—either to support or contest it—and how their different interests are at play throughout REDD's implementation.

CONTESTED MYTHS: DISGUISING AUTHORSHIP AND LIMITING PARTICIPATION

The idea that the SISA law was a product of civil society and public discussions was a crucial element in the government's myth of origin, providing legitimacy to the law and, crucially, reasserting Acre's authorship of the whole process. Many scholars have pointed out how elusive the idea of civil society can be, while also recognizing its increased currency,[25] especially in more recent contexts in which new modalities of government that rely on individual responsibility and forms of self-government emerge (Ferguson and Gupta 2002). These authors have called attention to civil society's Eurocentric origin and the false opposition between civil society and the state (Ferguson and Gupta 2002; Comaroff and Comaroff 1999), noting that what is commonly referred to as civil society is, in fact, made up of international organizations (Guyer 1994, 223). More specifically, Guyer argues, organizations of civil society "absolutely require state resources to exist" (226), while also "becoming as international as the states they engage with" (218). In other words, local or national organizations of civil society necessarily need to expand their networks internationally (even if only symbolically) to ensure their relevance. In the end, these international organizations, labeled as civil society, constitute an important element of transnational governance systems.

In the SISA consultation process, state authorities claim to have sent invitations to NGOs, scientists, governments, cooperation agencies, and companies at the local, national, and international levels (Governo do Estado Acre 2012, 18).[26] Three organizations sent their recommendations for the law in writing—WWF, Katoomba Group—Forest Trends, and Embrapa[27]—however, the international organizations IUCN, CCBA,[28] Cooperative for Assistance and Relief Everywhere (CARE), and EDF, as well as the national IPAM[29] and FGV,[30] and a few local NGOs also participated.[31] Both WWF and Forest Trends have been deeply involved in the implementation of the SISA law, whereby their authorship in the drafting of the law cannot be entirely dismissed. Therefore, the SISA process provides a good example of state functions being outsourced to NGOs (Ferguson and Gupta 2002, 989), even if that outsourcing is operating under the beneficial idea of a participatory consultation process. As such, while there is enough evidence of a broad involvement of international organizations in the process that culminated in the SISA law, local authorities need to claim a native ownership—not just to endow REDD with legitimacy and promote people's enrollment in the initiative, but also to ensure their own legitimacy as the government of the state. Ultimately, Acre's authorities have resolved this conundrum by leveraging the ambiguous concept of civil society—which encompasses international, national, and local NGOs—while truncating the de facto participatory process to exclude certain local actors, mostly those vocally opposing REDD.

The generation of value and what is perceived as valuable are equally contested issues within the law. While authorities carefully insist that SISA is not about creating a market for environmental services, but rather a mechanism for providing incentives for sustainability, their critics see officials as attempting to rhetorically disguise the commoditization of nature and new forms of green capitalism. Once again, this conflict is reconciled discursively through a filter. According to Roberto, a member of the state government who has been deeply involved in the whole REDD process,

> REDD is strictly a financial market. But what I think is Acre's geniality is that from the door out we recognize it is a market. So we worked on that binomial of understanding how the market works, but at the same time, making it in

a way that when value is generated, from the door in we have some sort of filter, a filter that takes into consideration the local reality and the interests of Acrean society.[32]

Roberto was thus explaining how Acre's authorities (including himself) were able to create a law that serves two distinct audiences and meets their different goals: on one hand, it enables the potential creation of a market, thus responding to international donors, or what he called "the door out"; but on the other hand, the law also includes providing local farmers and producers with state subsidies in return for certain practices—those subsidies (or incentives) appeal to their constituents, or what he called "the door in." But even if the filter that Roberto talked about can transform a future market for environmental services into some sort of "state social welfare provision" (Greenleaf 2016), what the SISA law seems to be is an attempt at managing the difficult balance between international donors' ambition of creating an international REDD market and local circumstances dictating the need to provide resources to local producers, while also avoiding any market-related language that can challenge the legitimacy of state policies or of the government itself.

For the "No REDDs," however, this balance operated by state authorities rendering market payments into subsidies only serves the purpose of buying people out, while falling short of being a real source of income.

> These offers [government grants], they are meant to alienate. Because if the minimum wage in our country is not enough for a person to live with dignity,[33] imagine a grant that equals, for example, the Green Grant, which is the one suggested in this REDD issue, it's a grant of R$300 every 90 days. Therefore, it's a way of inducing people to accept these policies.[34]

Daniela, the union member who made these remarks about REDD, is one of the strongest opponents of the SISA law and other policies carried out by the government in rural areas. Having actively participated in the rubber tappers' struggles, alongside Chico Mendes, against the expansion of cattle ranching in the state of Acre, Daniela sees schemes like REDD as a betrayal of that struggle and of her own identity as a union member. Unlike many political figures who claim to have been close to Chico Mendes and his ideals, Daniela is very discreet about that past, and does not mobilize it to assert her legitimacy as a union member. Her relentlessness (referred to by her opponents as stubbornness) in the defense of Chico Mendes's legacy is what has granted her that legitimacy among rural workers. Although her comment conflated a federal policy called Green Grant[35] with the SISA law, the mechanisms at play are, to her, the same. In both cases, what is at stake is the low income that such schemes provide, and the forms of dependency that they produce, which ultimately only alienate people.[36] During the same conversation, Daniela explained her opposition toward REDD: "It takes away from these people [people living in the forest] the right to live, to use nature in a way that is substantial." She was referring to the criminalization of forest dwellers promoted by schemes like REDD (when they cut trees down for their own use) but also to schemes like sustainable forest management that threaten the subsistence of forest dwellers due to the permanent loss of trees.

THE INTERNATIONAL VERSION OF THE MYTH: "A SUCCESS"

Despite the strong polarization around the SISA law and the contentiousness of the state government's plans for Acre's forests, the law has been deemed a success—especially among international environmental NGOs, but also by its donors. In a document introduced to the UNFCCC, the German cooperation agency referred to Acre as their "first early mover" in the implementation of REDD, classifying it an "early success in reducing deforestation through implementation of strategies for sustainable land use . . . including support for forest protection and indigenous communities" (Pfeil, n.d.). The REDD Early Movers program, established by the German Cooperation Agency (GIZ)[37] with the support of the German Development Bank (KfW)[38] has been one of Acre's main sources of funding, with €56.5 million pledged for the period of 2012–2019 in accordance with demonstrated results in reducing deforestation, plus €8.5 million for technical support. Norway is, however, the biggest donor for Brazil in general, and Acre in particular.[39] Despite the amounts involved in the creation and implementation of this law, Roberto put things in perspective for me during one of our conversations. According to him, Germany, "the richest country in Europe," gave Acre R$60 million, and Norway, "a great and excellent partner," provided, through the Amazon Fund, R$1 billion. In order to implement sustainable development policies, he continued, the government of Acre has spent, so far, R$2.5 billion—"two and a half times what Norway spent!" he emphasized, concluding that despite donors' good intentions, implementing sustainable development costs a lot of money.[40]

Despite the international claims of Acre's success, Roberto was well aware of the difficulties in implementing an international template while also making it look like a local design to address local problems. He was very well aware, too, of the need to produce results that could satisfy international donors while at the same time respond to the needs of Acre's population. Finally, as it became clear throughout some of our conversations, all decisions made by the state government had to also be carefully managed with Brazil's federal authorities, since states' autonomy can vary greatly depending on what is at stake. The creation and subsequent implementation of the SISA law embodied, thus, the effort of Acre's authorities to manage the difficult balance of governing locally but simultaneously engaging national and international actors and their interests.

From Myth to Practice: The SISA Law

The most tangible element of the SISA law is the governing structure that was created within the existing state government, which would be responsible for the implementation and regulation of the other elements of the law. I became aware that the specific organic structure that Acre authorities created responded to some demands of international donors (namely Germany) and followed an international template, which included a potential international market for carbon offsets. However, and despite the fact that state authorities seemed to have followed a predetermined template, the structure created was mostly informal.

The SISA governing structure includes the Institute of Climate Change (IMC) that regulates, controls, and registers any actions aiming at fulfilling any of the goals established by the law; a commission that validates and follows such actions (CEVA);[41] a scientific committee; a public-private company for the development of environmental services (CDSA);[42] and an ombudsman. Also connected to this governing structure is the State Secretariat of Forest Development, Industry, Commerce, and Sustainable Services (SEDENS), and the state Land Institute (ITERACRE), both of which preceded the SISA law. Except for the Institute of Climate Change and the CDSA, none of the other governing structures had a physical location, nor were their elements clearly identified in the documents available to inform the public about the SISA law. In order to identify the people responsible for the validation commission, the scientific committee, or the ombudsman, I had to ask specifically who these people were and where to find them.[43] And by 2015, despite the fact that the law had been approved five years earlier, the ombudsman's office had only recently been created, and it still lacked the necessary resources to carry out its functions. In fact, during a conversation with the two members of the ombudsman office, they recognized that although there were some reported problems with one of the private REDD projects in the state, their office lacked the resources to undertake a proper assessment of the situation, which would entail a trip to the project. While I was in Acre, this trip never took place.

It was apparent that while international templates suggest the creation of all these different organs, on a local level, the implementation of such a template was ruled by informality. People had been nominated for these organs, but resources were never allocated, either to create formal offices or to keep them on a permanent and exclusive basis. Resources were not enough to even allow them to perform their functions. All the elements of CEVA to whom I talked mentioned that they are not paid at all to exercise their functions; the same with the two members of the ombudsman. Inside the IMC, I met one staff member whose salary was being paid by WWF's local office, while the translations of all documents produced by IMC to be sent abroad (either for donors or as part of outreach material) were done by a freelance worker.

Ultimately, the governing structure of the SISA law is composed of a network of people who were already part of the state government or had worked closely with it in previous projects related to forest management. During my research, this structure suffered various shifts in the appointments, but overall, personnel remained the same—they just switched their positions from one office into another.

A notable exception to this already existent network of people was the aforementioned officer inside the IMC, supported by the WWF. This officer, of German nationality and with many years of experience working in Acre for the German cooperation agency GIZ, was responsible for the safeguards of the ISA Carbono program. He worked at the IMC, alongside other governmental workers and responding to the director of the institute, who is a state government appointee, yet he was not part of the state government payroll. One of the tasks of this person was to make sure that the documents produced by the state government matched the transnational requirements and templates. Given his experience and language proficiency, he was ensuring that state policymaking followed international procedures and coded language. The situation of this officer constitutes a telling example of the outsourcing of state functions to NGOs (Ferguson and Gupta 2002). More importantly, it demonstrates how by providing resources and technical support, international donors and organizations actually open the state policy apparatus to transnational interventions. Issues of language are, once again, crucial as the need to follow international templates and the appropriate jargon (be it in English or not) constitute important opportunities to open local and national governments to the support of foreign actors.[44]

In this context, even though the SISA law and its governing structure is presented as the legal outcome of public discussions about Acre's experiences and needs, it can actually be paralleled with other development projects, emerging from donors' priorities, modeled according to international templates, shaped and reinterpreted by practitioners' needs and local policymakers' circumstances—and eventually terminated once the funds are spent (or donors' interests shift to other projects) (Mosse 2005). The funds disbursed so far by international donors were able to determine the creation of a policy framework and corresponding governing structure in Acre—according to an international template—yet this framework was shaped by local circumstances, relying on a previously existing network of people.

Although Acre's officials insist on the nonmarket features of the SISA law, its template actually enables future market approaches. However, more important than the potential creation of REDD (or other ecosystem) markets is the subjectification effects that such institutional and legal frameworks can produce. In other words, even if markets are not created, the perspective that such markets can exist and be effective in the resolution of environmental problems becomes part of people's understanding of the world. As such, even if nonperformative in its stated goals, market-based mechanisms (such as those embedded in the SISA law) are culturally and politically performative in producing disciplining effects and disseminating neoliberal rationalities among conservationists and policymakers (Dempsey and Suarez 2016). In the end, despite presenting some difficulties to actually performing the tasks that were legally inscribed in it, the SISA law is already considered a success by donors, who continue to fund the program, and by Acre's authorities, who get additional funds, and see their state being championed internationally as a showcase for a unique jurisdictional REDD program in the world.[45] No matter how doubtful, such success claims are crucial for the law's legitimacy and, ultimately, for its survival (Büscher 2013).

MAKING SUCCESS BY DOING WHAT IS ALREADY BEING DONE: ISA CARBONO

According to the law that created the SISA, the system is designed

> to stimulate the maintenance and expansion of the following ecosystem products and services: the sequestration, conservation, maintenance and increase of carbon stocks and the reduction of carbon fluxes; the conservation of natural scenic beauty; the conservation of socio-biodiversity; the conservation of water and hydric services; the regulation of climate; the cultural valuation of traditional and ecosystemic knowledge; and the conservation and enhancement of soils. (Governo do Estado Acre 2010, 3)

While the text never mentions markets, the language used is informed by a market-based perspective expressed in words such as "products," "services," "stocks," or "valuation." In contrast to the vagueness of, for instance, what constitutes "natural scenic beauty," the text is straightforward on the system's governance structure, including its internal hierarchies and respective wages. It becomes evident that the specific object of governance (e.g., the definition of "ecosystem products and services") is vaguer than the determination of the institutional framework that will operate such governance. This results from the fact that laws like SISA derive from international templates that tend to be generic and standardized. Thus, while these types of templates—like the several studies published by The Economics of Ecosystems and Biodiversity (TEEB)[46]—refer to the need to give a price to specific ecosystem services so they can be preserved, they never specify how to do so, or even how to individualize those services into a commodifiable asset (Castree 2003, 280).

As drafted, the SISA law intends to develop all types of ecosystem services, but according to my government interlocutors, the issue of carbon was the most developed in their discussions, and that is why SISA started its implementation through the reduction of carbon emissions, the ISA Carbono.

And indeed, within the law, the section dedicated to the ISA Carbono is the longest, with seven articles, while all the other ecosystem services have only one article each pointing to the drafting of a future law to regulate them. Thus, although the law refers to environmental services in general, what it regulates is the potential to create carbon offsets through the reduction of deforestation.

Acre's authorities also recognized that, unlike other ecosystem services, the generation of carbon offsets through reduced deforestation allowed them to incorporate ISA Carbono in existing policies. In other words, this ecosystem service allowed them to continue to do what they were already doing, while obtaining additional funds to sustain those policies. They did not have to create any new program, which was openly admitted by Márcia:

> The program is practically based on our model of economic development. Therefore, having that in consideration, we are not looking for a partner to implement a project in a certain area, but a partner to develop the actions that we have been doing, such as the promotion of sustainable productive practices, or starting to develop a market for those products.[47]

And by Roberto:

> The state of Acre has this idea of REDD as leverage of public policies, of other public policies. So the idea is to potentialize existing public policies, from fish farming, forest plantations, forest replantation, production of sustainable energy, production of rubber,[48] we always try to pick up what we have already done, to understand what worked, and use REDD as leverage for those public policies that are working, or that we imagine will work in the future.[49]

The problem with Márcia and Roberto's perspective is that in terms of carbon accounting, it is fundamentally flawed. Given that one of the main rules of carbon accounting is that of "additionality" (meaning that all actions to reduce emissions must be additional to an existing scenario), the authorities' statements that the purpose of the SISA law "is to potentialize existing public policies" raises an interesting paradox: how are emissions being reduced additionally if those reductions are based on activities that were already in place? It could be argued that having funds to strengthen and

Table 2 Annual variation of deforestation rates in Acre

Years	Acre
2010/2011	+8%
2011/2012	+9%
2012/2013	-28%
2013/2014	+40%
2014/2015	-15%
2015/2016	+41%
2016/2017	-31%

Source: PRODES (http://www.obt.inpe.br/prodes/dashboard/prodes-rates.html)

expand existing policies can be a form of adding up, but it is hard to access the effects of the SISA law on deforestation rates, since these have neither been reduced on a consistent basis since the implementation of the law, nor are the reduction rates higher than the increased deforestation rates. Even if the 28 percent decrease in deforestation registered in 2012/2013 could be seen as a result of SISA's approval in 2010, such reduction was followed by a sharp increase in the deforestation rates, repeated again in 2015/2016, thus rendering it difficult to claim that SISA is having a positive (and consistent) effect in preventing deforestation (see table 2).

I found this paradox in Mozambique as well. However, in both Acre and Mozambique's cases, this issue was never raised as a problem, or at least as a potential factor to consider, in the official carbon data of these two territories.

The articles in the law referring to carbon identify the goal of reducing carbon emissions and enhancing carbon stocks (to be achieved through several actions, including sustainable forest management) and regulating the procedures to account for those actions (eventually including private projects), all in conformity with national and international rules. The national and international levels are indeed referred to five times throughout seven articles.

The ISA Carbono has as its specific goals:

> III—to strengthen the cooperation and alignment at the international, national, subnational and local levels. . . .

> V—to promote the institutionalization of a REDD+ state system based on nationally and internationally recognized concepts. . . .
>
> The [emissions reduction] units not able to be pre-registered . . . can be . . . used . . . in the fulfillment of reducing emissions programs or goals as a result of national policies and international commitments to fight climate change. . . .
>
> Article 26. The information contained in the state registry . . . are of public nature and can be sent to the national and international institutions responsible for the accounting of emission reductions . . . within the scope of national policies and international agreements on climate change and environmental services (Governo do Estado Acre 2010, 13–14).

While such references to the national and international levels can be easily understood in the context of a state policy that cannot go against federal legislation, such careful drafting can also be interpreted as an attempt at fulfilling two purposes: on one side, to use international commitments assumed by Brazil's government as a source of legitimacy for ISA Carbono; on the other side, to use an open language that allows for ISA Carbono to adapt to any national policy that the federal authorities might approve in the future.

The absence of a federal legal framework for REDD was, at the time of my first visit to Acre in 2013, a source of anxiety for both NGO practitioners and state authorities. Some of the practitioners I talked to during my time in Brasília manifested their discomfort with Brazil's lateness in approving a federal framework that could incorporate all the different state initiatives that were already taking place, such as those in Acre. However, the representative of the Ministry of the Environment claimed that Brazil did not need such a law, because the Amazon Fund and the recently approved Forest Code was Brazil's REDD.[50] This claim was clearly stated in Brazil's submission to the UNFCCC stating that REDD cannot be implemented as a series of stand-alone projects generating carbon offsets to be traded in international markets, and that, instead, it needs to include a set of wider public policies related to forest management that, altogether, can result in the reduction of carbon emissions through avoided deforestation.

Roberto confessed to me that state authorities would indeed feel more comfortable if the federal government approved a national REDD law—although he pointed out that the SISA law had been drafted in a way that enables its adaptation to any future legislation. But at the same time, he boasted about the fact that the federal Forest Code, approved in 2012,[51] had been inspired by the SISA law—"hence, Acre's pioneering!" he concluded. The Forest Code does mention the establishment of programs for incentivizing environmental conservation, including payments for ecosystem services. All the services identified in the law are the same ones mentioned by Acre's law 2.308 of October 22, 2010 (also known as SISA law), which, for Roberto, meant that federal authorities had taken inspiration from Acre's law. However, this list of ecosystem services is also part of what I call international templates, but with some local variations (see Wunder 2005).

Regardless of the validity of Roberto's claim about Acre's pioneering, the possibility of having the federal government challenge international transactions of carbon offsets generated by Acre's forests was broadly recognized, even if considered to be unlikely:

> Due to the lack of an international agreement on the climate that includes an international REDD system, and without the definition of a national framework on this issue, the legal base for the alienation of carbon credits by states like Acre can be juridically questioned.
>
> However, the states[52] ordered a legal opinion that justifies their rights over that public good [i.e., carbon credits]. (WWF-Brasil 2013, 68)

This potential legal conflict between federal and state jurisdictions would be later settled, when Brazil submitted its Intended Nationally Determined Contribution (INDC)[53] to the Paris COP in 2015:

> Brazil emphasizes that any transfer of units resulting from mitigation outcomes achieved in the Brazilian territory will be subject to prior and formal consent by the Federal Government.
>
> Brazil will not recognize the use by other Parties of any units resulting from mitigation outcomes achieved in the Brazilian territory that have been acquired through any mechanism, instrument or arrangement established outside the Convention, its Kyoto Protocol or its Paris agreement. (Federative Republic of Brazil 2015)

Brazil's submission to the UNFCCC was received with a feeling of vindication by those opposing REDD in Acre. One of my interlocutors even referred to the document as having a "demolishing" effect over the state government's intentions, since it precluded the sale of any offsets generated in Acre to international buyers. According to Brazil's submission, rates of deforestation reduction are to be accounted for at the national level and, therefore, the state of Acre cannot benefit financially from those reductions—at least not without the mediation of federal authorities. When I questioned authorities about this, state officials downplayed the demolishing effect, stating instead that, at least from that point on and from a legal standpoint, they knew what to expect from the federal government.

The REDD process in Acre provides a compelling example of how transnational governance operates simultaneously at the local, national, and international levels, and how national authorities exercise and negotiate their political agency in the interstices of these three levels of pressure. Moreover, events in Acre demonstrate that transnational policies are deeply contingent to local circumstances: although REDD is purportedly an environmental mechanism to reduce deforestation and carbon emissions, the reasons for Acre's authorities to enroll in it were not necessarily environmental. Instead, Acre's authorities sought opportunities to increase the economic growth of the state. The REDD template thus appeared to provide access to a significant amount of international funds that could be used to develop Acre through the implementation of "sustainable" and "productive practices" (which I detail in the next chapter).

International donors and other international actors are well aware of the potential different goals sought by local authorities when taking upon the REDD process. But that difference does not constitute a problem because, ultimately, what matters is the creation of an institutional apparatus that can be later used to enable and expand transnational governance. In the same way, it is not problematic that the SISA governing structure is so informal, lacking the needed resources to perform its stated tasks. Such informality is rather productive by easily allowing the incorporation of foreign elements, which can help to adequate the structure to international templates while also mastering the linguistic skills required in international arenas.

In the next section, I turn to Mozambique and continue to explore the productive intersection of different governance levels with environmental NGOs in the country's policymaking. As in Acre, in Mozambique the REDD process has produced an environmentality apparatus that enables the expansion of transnational governance.

Illegal Trade of Carbon: Myths and Realities

When I started my fieldwork in Mozambique in the summer of 2014, I was aware of the N'hambita project, which could be considered the first REDD activity in the country. It was a carbon offsetting project, implemented by the UK company Envirotrade, in partnership with the University of Edinburgh, in the buffer zones of the Gorongosa and the Marromeu National Parks, in the Sofala province.[54] Government officials and consultants in Mozambique whom I asked about this project were often ambiguous in their assessments, recognizing that, although private REDD projects were welcomed in the country, N'hambita could not be considered a successful initiative. Some did not consider it a REDD project at all. One curious and significant story related to this project was told to me several times: during an important meeting with members of the government, one official was explaining carbon's economic potential to the others, telling them that Mozambique could profit by trading carbon in international markets. This official mentioned the N'hambita project as an example, referring to the carbon offsets that Envirotrade sells across the world. His comments were quickly interrupted by a customs officer, a military man, who asked, "They say the carbon was exported to Mauritius, but through which border was this carbon transported?" His question was sincere and constituted a form of warning, suggesting that authorities had to act quickly to protect the borders, so as to make sure that this carbon was not being illegally transported out of the country.[55] The telling was usually followed by laughter.

Whether apocryphal or not, the story acquired mythological features and was always recounted with the purpose of demonstrating some of the fundamental challenges with efforts to implement REDD in Mozambique.[56] In particular, it emphasized how older government officials were confused by this sophisticated form of forest conservation. Not only did they not understand what REDD was, but they were troubled by the intangibility of carbon as a commodity—inasmuch as they referred to it as something whose movement across borders could and should be stopped. Although this narrative always implicitly indexed the "underdevelopment" of Mozambique, it can also be read as reflecting the anxieties associated with being at the periphery of the world system (Wallerstein 2004). Like so many other countries on the periphery, Mozambique has been a supplier of natural resources and raw materials for centuries, bound to a system that reproduces deep global inequalities. Hence, perhaps the earnest concern of the customs officer in trying to make sure that this new commodity would benefit Mozambique.

If the more explicit stereotyping this story invokes should be challenged because of its overtly ethnocentric tone, its subtle content perhaps bears deeper reflection in light of the new forms of transnational governance and subjectification that carbon trading entails. In some ways, this story points to how the new forms of transnational governance enabled by carbon trading rely, in large part, on local actors' lack of understanding regarding the commoditization of carbon.

INSTITUTIONAL (AND GENERATIONAL) RIFT

Ideas about how Mozambique could best implement and benefit from REDD have changed considerably since I first heard this story, but the inherent rift between an older generation of government officials (still marked by the experience of the civil war and foreign interventions in the country) and a new generation (more attuned to the rules of international trade and financial markets—including carbon markets) has remained. This rift was frequently described to me as an opposition between those who still saw Mozambican forests as having economic potential (through logging) and a more modern group of people concerned with the sustainability of these forests. But the disagreements were not so much about how forests were to be managed as they were about policymaking—or more precisely, who would have the power to make (and thus benefit from) policy. When I first started my fieldwork in Mozambique, this rift could be represented by the disagreements between the Ministry of Agriculture (MINAG) and Ministry of Coordination of Environmental Affairs (MICOA), the two ministries then responsible for the implementation of REDD.[57] While the latter was responsible for international negotiations with the several countries and entities promoting REDD, the Land and Forests Directorate inside the Ministry of Agriculture led the technical part of the program.

During the interviews I conducted with officials from both ministries, their differences became apparent both in terms of their perspectives about REDD and their technical preparedness. While people inside the directorate were very cautious about what REDD might entail for a country like Mozambique, inside MICOA, REDD was talked about with enthusiasm; it was, as officials frequently put it, a "good business opportunity." People inside the directorate were mostly "forest people" (forestry engineers with long careers connected to rural and forest issues and much experience in previous development projects implemented in the country). Consequently, during our conversations, they raised specific concerns related to land rights versus carbon rights, appropriate tree species to plant and/or preserve according to local conditions, and the need to move REDD slowly in order to not create false expectations, but mostly, to avoid potential conflicts. This cautionary note, whether or not it was substantiated by the realities of the field, reflected my interlocutors' memory of the war—or, at least, their greater sensitivity to the related political tensions still reverberating across Mozambique's landscape.[58] One of these officials explained to me that after the war, many people stayed in the bush, practicing subsistence agriculture and using forest resources for firewood and charcoal production. According to him, although the government considers these practices unsustainable, authorities are also very aware that they need to be careful when trying to convince people to change their livelihoods, since such efforts can be perceived as yet another unwanted interference by the government, leading people to get their weapons again.[59]

In contrast, my conversations with people inside the environment ministry (generally younger), always struck me as overly generic, as if these officials were just repeating some of the international jargon about forests and deforestation. Their discourses consistently neglected Mozambique's sociocultural and historical context, as well as the ecological specificities of Mozambican forests. This supposed lack of technical preparedness by these MICOA officers was referred to by one of the most outspoken local environmental NGOs, Justiça Ambiental (JA!), as purposeful, enabling the government to ally itself with business corporations and hurt people and the environment without any form of accountability. According to the two members of this NGO I talked to, the name of the ministry itself was telling: Coordination of Environmental Affairs.

> –They're just a coordination. They are not a Ministry.
> –Yeah, so they don't do anything. Basically, MICOA is there to make believe. Because they don't have the technical expertise—well, the technical expertise some of it they might have.
> –But they were set to be weak.[60]

The idea of a purposefully weak environment ministry in a developing country is not estranged to the fact that the international procurement of natural resources in the Global South is often facilitated by a consistent lack of legal enforcement of environmental legislation, and accountability by the companies exploiting such resources. The elements of JA! were precisely arguing that a weak environment ministry[61] served the purposes of international actors and local elites, especially in a context in which Mozambique was poised to become the world's largest coal exporter, is already an important oil producer, and has captured a lot of international investment since at least 2010, thanks to recently discovered offshore natural gas reserves.[62] Therefore, since my early days in Mozambique, it became clear that in the REDD process, the Ministry of Agriculture was almost limited to a technical role (that of carbon accounting), while MICOA seemed to be a mere advocate of a process whose leadership was very hard to pinpoint. In fact, both ministries' role appeared to be determined by the rules and conditions established by the World Bank within the REDD preparedness process, which defined the procedures to access REDD funds.

Beyond their different responsibilities with respect to the REDD process, I also got the impression that REDD meant something different to officials from each ministry. If, for the forest people inside MINAG, things had to start with the establishment of databases about forest resources, carbon stocks, and deforestation drivers, for MICOA, the process was all about drafting the documents required by the World Bank. These documents would not only entitle the country to obtain funds, but also enable more private foreign investments in the forest sector. In short, if for MICOA representatives REDD was an opportunity to capture investments, then for the folks inside MINAG, some of these investments could actually comprise veiled attempts at land grabbing. Mário, a MINAG forest engineer, vocalized his suspicions: "If carbon is being traded at $4 per ton (or even less), why is it that some investors want 8 million hectares?! It's not for the carbon," he added, "it's the land! They want big forest plantations and industrial agriculture."[63] Like Mário, other government officials inside this ministry were equally concerned about the possibility of triggering conflicts in regions where RENAMO has a lot of popular support, and therefore, where REDD projects can be easily perceived as another unwanted interference by the government that affects communities' access to forests and other resources (see box 2).

Box 2

Civil war in Mozambique

Right after Mozambique's independence, disgruntled Portuguese white settlers fled the country and joined the Rhodesian authorities with the intent of organizing an opposition force against the new Marxist government. RENAMO (Mozambique's national resistance) was thus created under the auspices of the Rhodesian secret services as a military unit with the goal of derailing the new state (Geffray 1991; McNamara and Meneses 2014) by sabotaging the economic infrastructure and the basis for the rural economy (Newitt 1995, 564).

FRELIMO's initial Marxist-Leninist affiliation informed its efforts to spread its presence and influence in the entire country with the purpose of eradicating all vestiges of colonial power. However, what FRELIMO considered to be unacceptable residues of colonial rule, like the *régulos*, for some rural populations these traditional chiefs represented legitimate leaders with strong spiritual connections to ancestors (Lubkemann 2008, 97). Consequently, FRELIMO became increasingly authoritarian, generating grievances that were skillfully leveraged by its opposition, especially during the late years of the Civil War (Geffray 1991).

Despite the vast network of support that RENAMO enjoyed—that included, besides Rhodesia, the apartheid regime of South Africa, and US secret services—its success was greatly indebted to the resistances and disaffection that FRELIMO's governance originated. Indeed, RENAMO was able to take advantage of the grievances of rural populations against FRELIMO's policies of forced "villagization" and attacks on traditional leadership, religion, and witchcraft. As the conflict became closer to a stalemate and the international pressures for a peace agreement increased, both parties reached a cease-fire agreement in October 1992.

FRELIMO leaders painfully realized their mistake of going against important social and cultural structures of the population and how that mistake had cost the alienation of the rural populations, especially from the center and north of the country. Despite their later efforts at recognizing traditional authorities as part of a strategy to reinforce the authority of the state by positioning them as mediators between peasants and the state (Florêncio 2005; Kyed 2007), RENAMO was nonetheless able to maintain strongholds of those areas up to this day.

More than a distinction between technical and political aspects of the REDD dossier, the tensions between these two ministries reflected both the fact that no one agency was responsible for the entire process and the blurriness of the process itself. This institutional rift was ultimately addressed after the 2014 elections. The new government restructured both ministries and their competencies, renaming them, and moving the sector of natural forests (as well as MINAG's forest people) from the Agriculture Ministry to the new Environment Ministry. More importantly, the government created a technical unit for REDD (the UT-REDD), granting it autonomy to act independently of both ministries, working instead with World Bank officials.

The members of this unit were government officers, but their hiring process had been coordinated by the bank, and it was never clear to me whether their salaries were paid by the government or the bank. The entanglement of this special unit with the bank was so apparent that even their offices were situated across the street from the bank's back gate. Noteworthy was also the fact, as I later learned, that many of the staffers in this unit, especially its leader Manuel, and his right-hand man Romeu, had both been employees in the N'hambita project. In fact, Romeu was the only person who emphatically defended that project against what he called an orchestrated campaign by some "No-REDDs" who did not like the project, despite "the good work that was being done."[64] The recycling of these staffers from the private sector into the government speaks to the lack of expertise in this specific field of carbon accounting, but also shows how the REDD network intersects the private and the public across its local, national, and international levels.

The creation of the UT-REDD, with its administrative and financial autonomy, simplified the REDD institutional landscape, but it also made more visible how profound the World Bank's leadership of the REDD process has been. Prior to this governmental restructuring, the role of the Bank was obfuscated by the several ministerial rifts, and lack of concreteness over what exactly REDD meant for Mozambique. It is that initial period in which REDD did not seem to have any tangible existence that I will now describe—this early period of the REDD process, and especially its institutional ambiguity, help clarify later events and how new modalities of transnational governance have been enabled in Mozambique through REDD.

Making the Country Ready: The "REDD Law" That Nobody Likes

As in Acre and many other developing countries, the beginning of REDD in Mozambique was a true transnational endeavor, both in terms of technical expertise and funding. In a country highly dependent on international aid, and where "the law itself is the 'object of development'" (Obarrio 2014, 35), it should not come as a surprise that a new mechanism for forest conservation came to be introduced through the same channels and means as development aid. The enthusiasm about REDD in Mozambique can be traced back to 2007–2008, when several NGOs and research organizations, including the American Woods Hole Research Centre, the Indonesian Center for International Forestry Research (CIFOR), and the British International Institute for Environment and Development (IIED) started to investigate the potential for developing these kinds of projects throughout Africa.

During the Copenhagen Conference of the Parties (in 2009), Japanese authorities selected Mozambique for a US$7 million grant to promote REDD. This grant included technical assistance for policymaking, acquisition of equipment, and provisions for technical cooperation with the Japan International Cooperation Agency (JICA) aimed at creating an information platform on "sustainable forest resources for REDD+ monitoring" over a five-year project (2013–2018).[65] This cooperation has since sustained the permanence of JICA staff in the Agriculture Ministry, inside the National Directorate for Lands and Forests (to develop this website platform), and funded visits by Mozambican officials to Japan (for capacity building and training sessions).

By 2009, Envirotrade Carbon Limited had already started the N'hambita carbon project, and the Norwegian company Green Resources had also initiated plans to generate emissions reductions for the Clean development mechanism[66] through planted forests in the province of Niassa.[67] None of these private projects were explicitly or exclusively defined as REDD (although the N'hambita project identified some of its components as REDD), but they increased local interest in carbon trading and pressed authorities to provide an adequate regulatory framework to avoid the "illegal transportation of carbon outside of the country." In other words, these projects, and the prospect of having other companies trying to develop similar activities (especially at a time when REDD was gaining momentum at the international level), motivated the government to regulate these companies' activities, while also adequately compensating the Mozambican state. Such regulation was quickly prepared and approved in 2013 in a "fast policy" process (Peck and Theodore 2015).

The REDD law, or "regulation of the procedures to approve Reduced Emissions from Deforestation and forest Degradation (REDD+) projects" was approved on December 20, 2013—at a record speed given the country's typical legislative timeline, which sometimes takes years. In nine pages comprising thirty-six articles, the law defines REDD; outlines procedures for approving projects, trading carbon credits, and consulting with communities; and clarifies safeguards, benefits sharing, fees, and penalties. It clearly states the fees that REDD project developers are required to pay, as well as stipulates annual fees (per hectare) and a 10 percent tax on each carbon credit traded.[68] This revenue is divided between the state budget (60 percent), the Environment Fund (20 percent), and local communities (20 percent).[69] The law is also very detailed on the several procedures a developer must satisfy to get any given project approved. Notably, there is only one article referring to the necessary consultations with the communities affected by such projects.

Although the law states that public consultations are mandatory and must precede the formal submission for approval, it is silent on how these consultations should proceed, what entities need to be involved, or even the minimum number of people who need to be consulted. There is also only one—equally vague—article on safeguards, stating the need to respect the existing laws on biodiversity and forests. In addition, it mentions the importance of respecting the rights of the communities (notably, their food security) and stresses the importance of allowing local communities to participate in the project design and implementation (República de Moçambique 2013, 1070). However, "rights" remain defined in very vague terms, without any specification on whether the law refers to land rights or the rights to the carbon of the forests where communities live. It did not take long for critiques of the REDD law to emerge.

According to JA!, the local NGO that opposes REDD the most, the law was pressured and rushed by the World Bank, and while some other bills have been waiting for more than four years to be approved, the whole REDD law process only took two. Even more shocking, only three months passed between the first public consultation about the decree and its approval to become a law, which makes it a remarkably fast policymaking process. As in Acre, the Mozambican REDD process sought its legitimacy through a public consultation process, while being pressured into a fast approval, ultimately justified by the urgency to reduce deforestation.

JA! opted out of this consulting process early on, as soon as it realized that the question of whether REDD should be implemented in Mozambique was not up for debate. That is, REDD was being implemented whether people liked it or not, and the public consultation on the law was no more than a means to legitimize the whole process through the mere presence of the consulted people—especially vocal NGOs like JA!. In this organization's eyes, the law fails to adequately address the rights of the communities and, substantiating their critiques, described several instances in which communities were duped into giving away their rights over land in exchange for promises (mostly unkept) of jobs, hospitals, and schools; as well as other cases of outright corruption, where local leaders agreed to give access to the communities' resources (like timber) to foreign investors in exchange for personal benefits—all cases that, according to JA!, were mediated by the government.[70]

Proponents of the law disagree. In a conversation with a legal Mozambican consultant who was involved in its drafting, I was told that there are not any risks to the communities because they will always have the final word on whether these projects can be implemented on their lands. According to him, the main safeguard rests on the fact that carbon rights do not necessarily coincide with land rights—that is, communities can maintain their rights to using the land while this same land is generating carbon offsets whose trading rights belong to a company.[71] In this perspective, the same land could simultaneously foster the material activities carried out by the communities and the intangible generation of carbon offsets—a commodity whose existence communities did not even need to be aware of. Of course, this entirely ignores the fact that carbon sequestration would by definition preclude certain types of land use—namely those that might involve deforestation. In a country where subsistence agriculture usually involves slash-and-burn techniques, this is at best a glaring oversight, and at worse, rhetorical nonsense. Moreover, this statement also suggests that somehow the very immateriality of carbon would (or should) immunize local communities from disputes that might arise over the attachment of new forms of remunerative value to land resources. This seems equally implausible.

The Mozambican constitution states that land is owned by the state and cannot be sold or taken by any other means. However, since 1997, laws opened the possibility of renting land, recognizing the right to use and harness it (Serra 2013, 59). This right is referred to by its acronym DUAT[72] and can be transferred between people by contract. Therefore, what this legal consultant was stating was that REDD projects do not necessarily entail the transfer of DUATs from communities to project developers, they just require the communities' authorization (see box 3). Another point neglected in his observations is that by law forest areas are always managed by the state, which has the sole right to attribute logging licenses. Given the fact that communities requesting such licenses are seldom successful, many opponents to the REDD law infer that communities submitting their own REDD projects would face similar obstacles.[73]

Box 3

Land tenure: Colonial continuities

Recent land deals in Mozambique bear strange resemblances to the ancient *prazo** system (Fairbairn 2013). Under the current land law, the state remains the sole proprietor of the land recognizing certain customary rights, thus allowing communities to use the land. However, efforts to attract foreign investments have prompted a de facto ownership of the land by private agents, as in cases in which commercial entities are allowed by the state to receive fifty-year land-use titles (which can be renewed for another fifty years) to develop certain projects (Walker 2012), such as natural resources extraction or industrial plantations. In certain cases, these land-use titles enable investors to maintain with the communities living in them the same types of relationships that the *prazos*' lords held over their *colonos*. That is, communities negotiate with private investors the use of their land plots or of their labor in exchange for infrastructure or other forms of development (Walker 2012). Continuities between older and current practices can also be noted in the nationalities of the investors involved in these land deals, which correspond to countries that have historically exercised power over Mozambican land, and that are currently also Mozambique's major donors, such as Sweden, Norway, Great Britain, South Africa, and Portugal (Fairbairn 2013, 336).

*The prazos consisted of gifts or concessions by African chiefs to the Portuguese and were one way through which the Portuguese effectively occupied and acquired jurisdictions in Africa (Newitt 1995, 218).

Notwithstanding these controversies, the law was drafted under the assumption that foreign investors—such as the ones involved in the N'hambita project—would be seeking to develop REDD projects in the Mozambican forests. However, after the law was approved, not a single project was submitted, nor was there evidence of interest by new investors. The law seemed to be useless by the time I first visited Mozambique in 2013, and two years later, the discomfort it generated among government officials was evident. Nobody liked it, and nobody knew what to do with it. Despite the allegations of an increased interest in Mozambican forests by foreign investors, some government officials like Natércia confided to me that what pressured them to create and approve the law at such a fast pace was the World Bank, which wanted Mozambique to join the international carbon market:

> Sometimes they [the World Bank] have been forcing us to be faster, and we believe that technically, we should not. Because we are not ready. Even if I, and other technicians are ready, the country is not. For instance, access to carbon markets is one of the questions about which we had very tough debates. Because we think we were not prepared for this yet.[74]

Despite the World Bank pressure to have a law ready, Natércia and others inside MINAG were concerned about the lack of knowledge about carbon and its trading among the population, not to mention the possibility of triggering conflicts over its revenues in rural communities. For Mário, the push for this law demonstrated how the World Bank is on the investors' side rather than Mozambique's. He believes that the bank's intention is to open the country to the voluntary carbon market even if that is not good for the country—as already experienced with projects such as the N'hambita.

The uselessness of the law was explained to me much later by Manuel, the head of the UT-REDD. For him, the law should have never been approved because it forces investors to pay too much money, discouraging them from doing something good for Mozambique. For him, the implementation of REDD projects can only be beneficial for the country, therefore, the law regulating these projects should encourage them, rather than charging fees:

> If you want to promote forest conservation, if you want to promote something good–how can you charge that person to do something good? On the contrary! The government should support, so you can help a community to do something good.[75]

Arnaldo, the World Bank team leader had a different explanation, pointing to a mistake in the evaluation of the situation in Mozambique. He said:

> In 2013 we came here on a mission, and then the vice-minister told us about this problem, that there were a bunch of REDD projects asking for permits, for the entire province, and they didn't know what to do. So I said, OK, let's start a regulating process, involving capacity building, and so on. So we spent almost a year preparing that law. And indeed, the law bends towards preparing the country for private projects. But in the meantime, what happened? Internationally, REDD completely changed the discourses. [International discourses on REDD shifted] out of projects into this jurisdictional thing. We no longer want projects. Nobody is interested in N'hambita anymore.[76]

Arnaldo justified, thus, the unsuccess of the law with the international shift of REDD from a project-scale into a jurisdictional one, like I had observed in Acre. Throughout my conversations with Arnaldo, and before him, with another World Bank officer in Washington, D.C., the role of the bank in the REDD process was always portrayed as merely following requests from the Mozambican government. Indeed, my first contact with the people from the World Bank involved in Mozambique was in D.C., at the bank's offices, since the Mozambican officer in Maputo was not authorized to talk to me without the presence of the D.C. focal point.[77] During that conversation, this officer—also a Brazilian national with experience in the Amazon region—kept deflecting my questions about the role of the bank in the policymaking process. Instead, she insisted on pointing to the information contained in the bank's Forest Carbon Partnership Facility (FCPF) website, noticing the "transparency" of the whole process, while repeating the mantra that "the World Bank does not have opinions" on what is the best for countries, and that "the World Bank does not choose which countries it will support" (countries ask for help instead). Significantly, and demonstrating how the networks created by REDD frequently intersect each other across international, national, and local levels, she told me, at the end of our conversation, how pleased she was about a recent "brown bag" event at the bank, where she had the chance to reconnect with folks from Brazil, namely Roberto, Danilo, and other individuals from Acre who traveled to D.C. to showcase the SISA law.

Part of the efficacy of this World Bank pressure lies precisely in denying its existence by either transfiguring the pressure into simple responses to requests of help or by effectively erasing the separation between the government and the bank. Such erasure is operated by the constant "we" deployed by Arnaldo when referring to the process of making the law—as if he was part of the government, and as if he had the competency to make policy decisions. Ultimately, the World Bank leverages the purposeful blurriness between national authorities' competencies and bank officers' initiatives to facilitate the bank's intervention in influencing the government's decisions. In the end, even if members of the government could attest to the pressure of the bank, its officers were still able to depoliticize the issue by portraying themselves as mere providers of help or by assuming a collective mistake in the assessment of the situation.

Legitimacy Through Similarity

The World Bank's leadership of the REDD process was reflected in a predetermined sequence of document drafting, fulfilling the bank's prerequisites for disbursing funds and continuing the procedures. The fact that the majority of these documents was not drafted by government officials or even by Mozambicans reflects how little ownership Mozambique had over the whole undertaking. The official documents requested by the World Bank's FCPF as part of Mozambique's application for its funds had been prepared by two Brazilian NGOs, one South African consulting company (with offices in Maputo) and two Portuguese ones connected through a consortium.[78] The preparation of these documents consistently ignored a historical repository of previous studies existent in the MINAG offices, not because they were being commissioned by a new entity—the UT-REDD—but because what mattered in these documents was their compliance with the World Bank's formulaic templates, not their content or the accuracy of their information.

Two such documents were the so-called R-PIN and the R-PP. In 2008, Mozambican authorities submitted the readiness plan idea note (R-PIN) to the World Bank—a twenty-page document that served as an application to access the funds provided by the bank's FCPF. It is unclear who exactly prepared this first document, but in all likelihood, it was drafted by a Brazilian NGO, with funds provided by Norway.[79] Approved by the bank a year later, this document can be seen as the starting point of REDD in Mozambique. After its approval by the FCPF, Mozambique obtained US$200,000 to prepare the readiness preparation proposal (R-PP),[80] which, after its approval, provided the country with US$3.4 million more to draft the next document, the national strategy for REDD+. Amid this document drafting process, the embassy of Norway promoted and funded a "South-South Cooperation" between Mozambique and Brazil. Hence, in March 2009, Mozambican authorities signed a memorandum of understanding with the Brazilian Amazonas Sustainable Foundation (FAS)[81] with the intention of promoting the Brazilian experience in implementing REDD. For Norway, I was told, this type of exchange made sense because the support given by this country to the Brazilian Amazon Fund as a precursor to REDD is considered successful, and therefore can and should be replicated in other locations as a means to expand REDD. Moreover, since both Brazil and Mozambique share the same language (Portuguese) the exchange of experiences would be easier.

The fact that most World Bank officers working in Mozambique were Brazilian nationals was not a coincidence. The World Bank seeks these linguistic affinities to facilitate the work of its officers in the countries where they operate. However, this bank procedure also nicely dovetailed with longstanding assumptions of Brazilian diplomacy, which deploys a language of culture to claim that relations between Brazil and African countries are special, and different from those developed by countries from the Global North (Cesarino 2017). According to this diplomatic rhetoric, such differences could be explained by the fact that both Brazil and African countries share cultural traits, and historical and geographical circumstances, which render Brazil a natural interlocutor for African nations. This argument has been explained by several scholars, who referred to it as a "culturalist grammar" (Saraiva 1993), a "similarity of circumstances" (Shankland and Gonçalves 2016), or the use of the "affinities discourse" (Cabral and Shankland 2013).[82]

The South-South cooperation promoted by the World Bank and Norwegian authorities perfectly matched diplomatic discourses about the relations between Brazil and Africa. But, as noted by Cesarino, this type of cultural diplomatic rhetoric is only deployed by the "upper" scales of diplomacy, while the "lower" scales (or the implementation side of diplomacy) do not mobilize that rhetoric at all or show little to no interest in cultural arguments (2017, 334). For the African counterparts, the cultural discourses of Brazilian diplomacy have little purchase too—instead, they are more concerned with "the technical knowledge and material resources that could be imparted by the Brazilians, or with networking with them for future opportunities" (Cesarino 2017, 335).

Nonetheless, the World Bank took upon this Brazilian diplomatic rhetoric, seeking to introduce REDD in Mozambique by indexing the mechanism to successful experiences in Brazil. Accordingly, such success could be replicated in a similar territory—Mozambique—thus providing legitimacy to what the bank was proposing. In these South-South cooperation workshops Brazilian practitioners from FAS claimed that Mozambique's deforestation problems were similar to the problems Brazilians faced, and therefore, the solutions adopted there could work in Mozambique too. In these exchanges, this supposed similarity became almost unquestionable, even though the tropical forests of Amazonia have little to do with Mozambique's miombo forests, and the deforestation drivers in each country are very distinct. In fact, upon my return to Mozambique, this similarity with Amazonia was openly refused by Arnaldo, who recognized that Mozambican forests had nothing to do with Amazonia tropical rainforests. Instead, he suggested, the Brazilian *cerrado* offered a more significant comparison. Tellingly, some of the development initiatives that were implemented in that area of Brazil to expand agricultural production are similar to the current plans that the World Bank and Mozambican authorities intend to develop in the province of Zambézia. Despite these misrecognitions, this cultural rhetoric was efficient to the extent that even though there was no longer any formal cooperation with Brazilians by the time I started my fieldwork, Brazil's perceived success in REDD was often mentioned as a good example to follow.[83]

Although funded by Norway and subscribed by the World Bank, this "South-South" rhetoric was also technically supported by the British International Institute for the Environment and Development (IIED), the University Eduardo Mondlane, the local NGO Centro Terra Viva (CTV), and the Finnish forestry consulting company INDUFOR. The involvement of these entities—all of them highly supportive of REDD—in the organization of this cooperation can be equated to what Cesarino called the upper scales of diplomacy. It was at this level that the enrollment in REDD as a mechanism to reduce deforestation in Mozambique was promoted and legitimized, through the organization of these workshops and the production of documents and promotion materials to inform and train lower-scale officers and potential future implementers. And if FAS, a small NGO located in the state of Amazonas, can hardly be considered part of an upper-scale diplomatic effort, its choice as the main interlocutor for the cultural exchanges on

Brazilian and Mozambican forests had a more pragmatic reason: Arnaldo, the World Bank team leader, already knew FAS founder and director, as well as other members of that NGO with whom he maintained personal ties. Indeed, one of my interlocutors in Maputo, deeply involved in the process of REDD's document drafting, once complained to me that Arnaldo kept bringing more Brazilians to Mozambique, and that they all seemed to be "buddies,"[84] a situation that underscored the importance of social networks in facilitating the work of NGOs and their funding agencies (see Hilhorst 2003, 24).

The short duration of this cooperation[85] still exposed Mozambican officials to the program Bolsa Floresta (Forest Grant), a pilot reward mechanism for environmental services in Brazil, described by FAS as a payment mechanism for environmental services. However, this program was modeled on a federal program of conditional cash handouts for poor families in Brazil, implemented with the goal of reducing poverty, the Bolsa Família (Family Grant).[86] The fact that Bolsa Floresta could be described as a market for ecosystem services, and therefore a form of implementing REDD, denotes its ambiguous nature, simultaneously championed as a market mechanism, while being implemented (in Amazonas) as a conditional-aid scheme. It also points to neoliberalism's polymorphism, whereby marketization can be combined with programs of direct distribution (Ferguson 2015).

For the Mozambicans exposed to this exchange, however, the Bolsa Floresta only demonstrated that while Mozambique depends on international donors (and therefore on their conditionalities), Brazil has the financial autonomy to do REDD the way they want—without pressure. The government officials I talked to about this exchange manifested their appreciation for the programs Bolsa Floresta and Bolsa Família, but at the same time, they also expressed their skepticism about the possibility of implementing similar experiments in Mozambique given the chronic lack of resources. What became clear from my conversations with these officials is that while the World Bank insisted on using Brazil as a successful example of REDD to legitimize its implementation in Mozambique, to Mozambican officials, the Brazilian experiments continued to be extraneous, thus inapplicable, in Mozambique. As such, what these officials knew at that point was that they had to prepare all the documents the World Bank required in order to continue accessing funds and fulfill the necessary steps to become "REDD ready."

Erasing Locality Through Document Drafting

The need to prepare these very specific documents required by the World Bank explains why FAS was chosen to be part of this South-South cooperation. FAS had considerable experience in the drafting of these types of documents, which made it appropriate to also work as a consultant for the Mozambican government. These documents conform to a certain genre, and therefore require meeting protocols, follow determinate templates, and perform according to a given style. FAS mastered the World Bank's language, and therefore began helping Mozambicans to draft one of the most important documents of the REDD-readiness phase, the national strategy for REDD+. This strategy is supposed to identify the main deforestation drivers of the country and enumerate all the activities and policies that a government is willing to implement to tackle deforestation and enhance carbon stocks.

During my first visit to the country, I learned that the strategy was still being prepared by scholars from the university who had been in training with FAS—even though such cooperation was no longer taking place—but it was unclear when the document was supposed to be ready. Later, I was told that the REDD national strategy that had been prepared by Mozambican scholars with FAS's contribution had been abandoned, and a new document was being drafted by a different Brazilian NGO hired by the World Bank and the UT-REDD. Despite my inquiries on why a whole new document was being drafted from scratch, the only apparent reason for the change was that the new document was being prepared in alignment with the rules required by the FCPF, the World Bank program. To my surprise, I found out that this new NGO was the same that had developed a popular REDD project in the Brazilian states of Rondônia and Mato Grosso involving indigenous people. In fact, I had already interviewed its founder, Celso, while still in Acre, given his connection to a Californian NGO that has been working with payments for ecosystem services among indigenous people across several Brazilian states, including Acre.[87]

COMPARING NATIONAL REDD STRATEGIES

The comparison between these two documents—the strategy that was still being completed in 2014 (República de Moçambique n.d.) and the one that was approved later, in 2016 (República de Moçambique 2016)—provides important insights into how transnational governance operates at the level of discourse. It is apparent how the problem of deforestation is differently depicted; how, accordingly, different solutions are identified; and how the document that was actually submitted to the bank carefully elided Mozambique's specificities while simultaneously erasing that elision process. If, in the old document, the problem of deforestation was presented as having multiple drivers and a complex relationship with forest degradation, the new one clearly identifies those drivers as shifting agriculture (carried out by rural populations in the absence of more advanced agriculture technologies), and charcoal production. The old document mentioned commercial agriculture, urbanization, development of (needed) infrastructures, and mining as important drivers, whereas the new strategy only briefly refers to these, completely ignoring mining activities (which are currently an important source of foreign investment).

Significantly, the previous document established that forest cover in Mozambique was 70 percent of the territory, while the new one dramatizes the problem, with only 50 percent of forest cover.[88] Most notably, the solutions to deforestation are also different. Given the complex portrait of deforestation and degradation drivers presented in the old document, the strategic goals included, among several measures, better land use planning policies and environmental assessment plans in the development of infrastructures, electrification of rural areas, incentives to the use of domestic coal (instead of charcoal), improvements in housing (to reduce the use of wood in the construction of precarious homes), elimination of wood concessions without management plans, and initiatives to better prevent wildfires. In short, a complex set of integrated measures that sought to include REDD in a broader developmental and environmental framework.

In contrast, the strategy that was approved makes no reference to any of these measures, and instead focuses far more on the role of conservation agriculture (to preclude shifting agriculture), more efficient methods to produce charcoal and distribution of improved cookstoves, strengthening the protection of conservation areas through ecotourism projects, and industrial plantations. In opposition to the old document's focus on rural communities—their rights and their role in forest management—the new strategy notices instead the opportunities brought by private companies, pointing to the need to create more beneficial fiscal policies.

Overall, it is clear that while the first strategy was focused on Mozambique's national circumstances, providing a detailed picture of the country's problems, its legal and institutional framework, and the history of its involvement with the REDD process, the final document is vaguer, more aligned with REDD's international template, and informed by an economistic language focused on a risk-versus-opportunities perspective, favoring markets and the private sector. Besides these two specific documents, I had access to others while they were still being prepared by various consulting companies hired by the World Bank. All these documents were commented on and edited by World Bank officials (or by other hired consultants) before being publicly released in their final format. In many of these side comments (at times inscribed by consultants who were never part of the several public consultations conducted in different provinces), it was very clear that the final documents were to be consistent with a simplified narrative on deforestation: its vectors are shifting agriculture and charcoal production.[89] Ultimately, and without much editing, all these documents about Mozambique could be applied to any other poor country with forests and a significant rural population. A more detailed analysis of these two documents shows, thus, that the World Bank *wanted* the drivers

Figure 7 Forum on REDD held in Quelimane, Zambézia; the UT-REDD and the World Bank introduce the new Landscape Approach to Zambézia. Photo taken by the author. Quelimane, 2016.

of deforestation to be shifting agriculture and charcoal production—despite evidence pointing to more complex and diverse scenarios, especially across Mozambique's different regions. Tellingly, shifting agriculture has been portrayed, since colonial times, as one of the reasons for Mozambique's lack of development (see box 4).

The simplified narrative about deforestation that the World Bank has consistently sought to maintain—one that ignores localities and focuses on slash-and-burn and other activities carried out by poor rural communities—is not just the outcome of a known logic inside the development industry. Scholars like Mosse (2005) and Ferguson (1994) have pointed out that "development" produces its problems in relation to the solutions that it can offer, ignoring structural causes or crucial interconnections with the object of intervention. In other words, the problems addressed by the development

---

Box 4

Shifting agriculture

At the end of the nineteenth century, Portuguese colonial authorities sought to modernize their colonial administration by replacing the semi-feudal obligations of the *prazo* by new capitalist ventures in the form of plantation concessionaires under chartered companies (Isaacman 1972, 161). These companies would henceforth be the means to establish an effective Portuguese occupation of Mozambican territories and promote industrial agriculture. However, by 1915 it was already clear that such ambitions had failed.

This failure can be explained by several facts, including neglecting the geographic, economic, and social realities of the Zambezi region, or the absence of real economic incentives for the companies to invest in large agricultural enterprises. Importantly, the indigenous pattern of shifting agriculture—a native strategy against the physical and geographical challenges of the region, and which had been already responsible for the failure of the *prazos* as agricultural endeavors—was also an important factor in the unsuccessful enterprise of the chartered companies to develop agriculture. Fast-forwarding to the present, shifting agriculture is yet again presented by the World Bank and the Mozambican government as one of the main culprits in the country's underdevelopment and rates of deforestation.

industry are consistently constructed as isolated from their historical and cultural contexts, disconnected from other problems, and depoliticized, thus solvable through the technical solutions offered, which are themselves utterly simplified.

Accordingly, it is easier for the World Bank to provide projects of conservation agriculture and improved cookstoves than to reduce deforestation by convincing national authorities to enforce more stringent environmental regulation (e.g., regarding major infrastructural investments on coal and oil), or by providing national authorities with the resources to enforce the legislation on illegal logging, or by addressing the structural conditions of poverty that increase the pressure over Mozambique's natural resources. Simultaneously, for national authorities it might seem easier to deal with poor rural communities (pressuring them to change their agricultural techniques and energy uses), than with corporations (many of them foreign), or with other actors linked to wider networks of international trade. But more importantly, this simplified narrative—just like the carbon-based simplified narrative about climate change—serves the goals of the countries from the Global North, donors, and investors of the World Bank. That is, this simplified narrative enables these countries to continue intervening in the rural world of the Global South, to continue pushing for schemes like REDD as the panacea for deforestation, and to develop new investments that, without the legitimizing discourse of saving the planet from climate change, would be more contentious than they already are.

All the document drafting that I encountered in Mozambique, and in particular the abandonment of the national strategy (prepared by Mozambican scholars) and its replacement by a new document, is a clear demonstration of the lack of ownership by Mozambican authorities of the REDD process. Even when the ministries held previous studies on Mozambique's forests and rural populations, the World Bank demanded that new documents were produced and that foreign consultants were hired with that purpose. As in Acre, the writing of specific documents in response to the demands of international actors (be they donors or the World Bank) opens an important space for transnational governance. Under the argument that Mozambicans lack the skills to fulfill the formulaic templates of these documents, foreign "experts" are hired to work either inside the ministries and governmental offices, or outside, replacing the competencies of the government.

Significantly, the writing of these new documents and studies is also an important part of the process of constructing legitimizing narratives—not for local Mozambicans, but for international audiences. For these actors, what matters most in these narratives is the depiction and disguise of the real causes for deforestation (disguising them with slash-and-burn and the use of charcoal) or opportunities to invest in the extractive sector of poor countries. By erasing localities and the specificities of the Mozambican forests, it is not just a standardization effect that is achieved but, more importantly, the legitimization of using the Global South as the privileged arena to reduce global emissions.

By 2014, REDD appeared to have little physical reality in Mozambique. There were multiple studies about the potential of the country and how to implement such projects, numerous workshops and training sessions with governmental officials, World Bank officers, and Brazilian NGO practitioners, as well as an equivalent number of documents circulating. Despite the absence of concrete activities on the ground that could be called REDD, Mozambican authorities were deeply involved in this new policy landscape.

The contrast between this rush of activity around REDD and the absence of concrete REDD projects was striking, especially given that everybody I talked to—either authorities or those opposing REDD—attributed a great importance to this new forest conservation mechanism. The words of a Norwegian diplomat I interviewed in Mozambique echoed throughout this scenario, as he declared, "REDD does not exist yet!"[90] But in the end, REDD had a real existence in the form of all these meetings, exchanges, and written papers. Although nothing concrete was being implemented at that time, all the rush around REDD committed Mozambican authorities to a new way of seeing carbon—as a commodity (not to be transported illegally out of the country)—and forests as a new potential source of funds through foreign investments. All the rush of activity around REDD was thus part of the implementation of disciplinary practices of globalized environmentalism (Goldman 2001, 503). More important than having any actual REDD initiatives on the ground, the rush of workshops, training activities, and document drafting and circulation prepared Mozambique for further transnational interventions, not just through the subjectification produced by these workshops and documents (i.e., fostering new perspectives on carbon and forests), but also through the opening of the governmental structure to a new governance apparatus in the form of specific units, consultants, studies, and ultimately, laws and policies.

The apparently messy landscape of REDD in Mozambique—with ministerial rifts and tensions, disagreements over what constitutes a REDD project or not, lack of clear ideas about what REDD can be in Mozambique, blurry boundaries between the competencies of government officials and private consultants, and even the creation of laws whose applicability remains doubtful—constitutes an important part of the REDD implementation process. On one hand, such messiness reflects how local, national, and international actors negotiate their role and agency in the pursuit of their own goals within this vast REDD network; on the other hand, it is precisely that messiness that facilitates the intrusion and expansion of transnational governance into Mozambique's political landscape by rendering such governance almost invisible. In the end, the World Bank is only responding to the requests of the Mozambican government for help to implement REDD.

Conclusion: Expanding Transnational Governance

Even if very different, the experiences of Acre and Mozambique share some common features. In both cases, the REDD process was pressured by international actors and marked by a remarkably fast policymaking procedure. This pressure was exercised—constantly and unevenly—at the intersections of the international, national, and local levels in the pursuit of different (and, at times, contradictory) goals by the multiple actors involved in the REDD network, ultimately enabling the expansion of transnational governance. The tensions between the Brazilian government and authorities in Acre (spurred by international donors and NGOs), as well as the blurring between the competencies of the World Bank and those of the Mozambican government, are good examples of this continuous pressure.

While international actors pressured local and national authorities to facilitate the construction of a policy apparatus, its creation does not mean that local and national authorities can be discarded henceforth. Rather, transnational governance is still dependent on enabling moves by such authorities. Such moves can be the mere compliance to prepare certain documents according to predetermined templates, the creation of specific governing structures, the formulation of laws, or the subordination of local knowledges to the knowledge produced by international experts and consultants. Indeed, what this whole process made possible in Acre and Mozambique was the application of an "environmentality" apparatus (Agrawal 2005; Agamben 2009) that, following the templates given by the UNFCCC; the World Bank; development agencies; international NGOs like WWF, VCS, or Forest Trends; and foreign donors, produces subjectification effects (Agamben 2009, 19–20). Such effects are visible, for instance, in the language used by authorities when referring to the problem of deforestation, notably when considering small farmers as the culprits of deforestation. They are visible too in the successful introduction of payments for ecosystem services in Brazil's Forest Code. Even if disguised in a language that emphasizes incentives and eschews references to markets per se, the market-based approach that Brazilian federal authorities have consistently resisted throughout the UNFCCC negotiations is now enshrined in a national law, the Forest Code. Although this legislation refuses carbon trading, it implicitly inscribes that possibility, and by doing so, it is adopting a new perspective on Brazilian forests that did not exist before. Similarly, in Mozambique, a country that maintains a communal approach to land rights, authorities were pressured into quickly approving a law that not only challenges the rights of the commons, but also implements a market approach to carbon, with implications over communities' access to land.

In this subjectification process, NGOs played an important role, providing discursive frames, stabilized knowledge, and legitimacy to the REDD processes (Li 2005). Big international NGOs like the WWF or EDF have been very active in pressuring for certain policies, not just at the international level where environmental policies are discussed (as in the UNFCCC), but also at national and local scales, promoting the implementation of policies enabling REDD. Some smaller NGOs have learned how to muster the international climate change jargon, underscoring the legal and policy apparatus needed to enable REDD and other forms of forest governance, operating thus, not just within their traditional jurisdictions, but also internationally—as in the case of FAS in Mozambique. Finally, local NGOs are equally important. Although frequently lacking funds and human resources, they are important sources of local legitimacy in the introduction of new policies, hence their frequent cooptation by international NGOs.[91] Despite the efforts to provide REDD with legitimacy and native origins, however, both processes have been highly contentious, generating criticism inside and outside the governmental structures.

The lack of institutional strength to lead such a complex process of capturing donors' funds, preparing documents and legislation, and capacity building was apparent in both cases, in tasks that were developed by NGOs, international donors, and hired consultants. Not surprisingly, neither Acrean nor Mozambican authorities hold authorship of their processes, even though the former invested significantly in producing a myth of origin that codes this process a product of Acre's "civil society." In Mozambique, the legitimacy for REDD has mostly been sought by the World Bank (not national authorities) using arguments of similarity regarding the Brazilian reality. The participation of outside experts in both these processes can be partially interpreted as part of neoliberalism's feature of transferring state functions to NGOs, but ultimately, it is also a crucial element of transnational governance. The intrusion of these elements in the existing governing structures of states—both in the form of staffers (a German in Acre, Brazilians in Mozambique) and the documents produced by them—produces the double effect of subjectification, and effectively opening those structures to transnational governance. And the less clear that intrusion, the more effective it is, even if that lack of clarity is not purposeful or intended.

If looked at from an environmental perspective, its explicit raison d'être, REDD can easily be seen as a failure, since it does not really reduce any emissions. As I established in chapter 1, REDD's supposed potential to cut emissions is totally dependent on fictitious assumptions over past and future scenarios of deforestation and misguided equivalences between forest carbon and fossil fuels carbon. However, that is beside the point. As unsuccessful as it can be, REDD still proved to be very useful for transnational governance, both in Acre and in Mozambique, working as a catalyst to the implementation of an environmentality apparatus.

This usefulness should not be interpreted, however, as if both of these processes were merely imposed (or pressured) from the top, forcing locals into embracing something that in the long term will most likely be damaging to the poor populations of these territories. Even if this transnational governance was able to create a legal framework that facilitates future interventions and expands market approaches in these regions, both Acrean and Mozambican authorities also took advantage of this opportunity. Both saw in REDD the chance to capture much-needed funds to pursue their own goals, be it those to maintain development policies in one of the poorest states of Brazil or to capture the interest of more private investors in Mozambique's rural areas. The long-term effects of these processes are still difficult to assess, but it is increasingly clear that the purpose of fighting deforestation has opened extensive rural areas (many of them outside of the scope of forest conservation) to the intervention of this transnational governance, as I will explore in chapter 5.

Chapter 5

REDD on the Ground

Broadening the Scope, Fostering Vagueness and Confusion

Introduction: Shifts in the International Template

Throughout my fieldwork in Acre and in Mozambique I realized that the activities that were being implemented under the REDD umbrella had changed considerably over time. If, during my initial fieldwork in these two sites, it had been immediately apparent that REDD as defined in the climate negotiations or even among staffers from NGOs based in Washington, D.C., looked very different from REDD on the ground, over time and over my repeated visits between 2013 and 2016, I also realized that REDD on the ground was itself subject to reformulation over the course of its implementation. It was not just that REDD's implementation had to meet local imperatives, as in other kinds of development intervention, but rather that REDD seemed capable of becoming almost anything that people needed it to be. It is perhaps this quality that makes REDD such a powerful catalyst of potentialities—some already existent, others as of yet only imaginable in the future. Indeed, REDD remains a difficult thing to define. However, it is perhaps this very elusiveness that enables it to open a considerable amount of space for negotiating and reframing what activities can be carried out in its name, how it can be implemented, what it can include or exclude, both at the international level (in abstract, as a template), and at the national and local levels (in practice). The plasticity of the REDD template has thus enabled not just the expansion of the scale of transnational governance (into jurisdictions) but, most importantly, of its scope, authorizing interventions that go well beyond the forest sector. This capillarity warranted by REDD explains why this mechanism continues to garner so much support, enabling the pursuit of many different goals by the multiple people involved in it.

The profound changes that REDD underwent since its appearance, and its resulting indeterminacy have led some scholars to anticipate its failure (DeShazo, Pandey, and Smith 2016), or at least its gradual transformation into something else that looks less like a market and more like conventional development (Fletcher et al. 2016). In this chapter I argue that the changes REDD has undergone should not be interpreted as a sign of either its failure as a policy nor a harbinger of its imminent demise, but rather as evidence of its potential. The acronym REDD may in fact be replaced by something else (such as the new epithet "landscape approach," for example), and yet, the core idea that it currently references—that we can reduce carbon emissions by reducing deforestation—retains a powerful grip in the policy imagination. Indeed, the Paris Agreement does not mention REDD anywhere, but the idea of using forests to reduce emissions is clearly warranted in the text. REDD's potential relies precisely in the vagueness of its definition and of its operationality on the ground (expressed in its capillarity), as well as in the legal and institutional confusion that its implementation often entails, opening the scope of transnational interventions.

The gradual changes and adaptations of REDD that I witnessed in Acre and Mozambique throughout my fieldwork, especially its expansion in scale and increasing capillarity, demonstrated REDD's plasticity. In examining these changes, I will show that despite the multiple reinterpretations of how to implement REDD, or of what REDD can be, and despite the various reasons to take up this mechanism, REDD continued to be productive—albeit in different ways—to local, national, and international actors. Ironically, REDD continued being productive and continued enrolling participation by allowing people to deny its very existence.

THIS IS THE POLICY'S "TIPPING POINT" MOMENT

> There is a moment, I don't know, the Acrean people are, at the time of election, it has been electing for 15 or 16 years a government that has been supporting this project. But because it is very hard to make this transition from an unsustainable model into a sustainable one, the last elections were

tight already. Very tight! Very tight. So you run the risk of, at some point, the Acrean society turn to you and say: "my friend, you're not doing your part, and we're hungry here." There will be a moment when you can't [continue with this model]. The economic model changes, the political majority changes, and you change the development models. People are going to do something else. People are going to make money, and leave. Take timber, put cattle, put soy—because people are making money that way. So, there is a limit, including a political limit, but also economic, a political limit of the growing capacity of this type of proposal. In the last election the government won with 50 points! So you have a divided society. And here it is very interesting because it is very clear this division between models. The political division clearly represents the economic division. So we are at the threshold of a choice. It's not easy. People need to see their lives improved. Either you accelerate the results, or the political model exhausts itself.[1]

When I sat down with Roberto in 2015, two years after I had started my work and we first met, his enthusiasm for the SISA program seemed to have dissipated. I had already learned that the structure created through the law had evolved significantly: the Institute of Climate Change (IMC) had a new office, as well as the Company for the Development of Environmental Services (CDSA), and the temporary appointments within the structure were now permanent, public consultations were completed and the state program had recently been certified by the Verified Carbon Standard (VCS).[2] Overall, things looked to be working the way they were supposed to, and thus, Roberto's despondency was unexpected. After all, Roberto had been deeply involved in the design and implementation of the SISA program when he held a temporary position in the IMC. When I first met him two years before, he not only knew every detail of the program, but also spoke enthusiastically about the way Acre had been able to adapt an international program to the local reality, about how SISA was a source of inspiration for federal legislators, as well as a showcase that could teach the world how to protect forests at the jurisdictional level. Why, then, only two years later did he seem so dismayed?

Perhaps the shift in Roberto's mood could be explained by how slowly things seemed to be moving—after all, the program had begun in 2010 with the approval of the SISA law, and in 2015 it was not yet fully implemented. Or maybe he was dismayed by the increasing protests against the state's program and the several private REDD projects taking place across the state.

Whatever the reasons for his feelings, as I continued to catch up with other people, it became clear that there was a different atmosphere surrounding the program: less celebratory, and more pragmatic. The several issues raised by the implementation of a jurisdictional program were taking a toll on state authorities' enthusiasm. One of the most contentious was the articulation of the state program with the several private projects already taking place on the ground—an articulation that is necessary to ensure that the carbon offsets generated by private projects are not doubly counted by state authorities.

The other important issue was related to land tenure, and the different legal jurisdiction levels that coexist in a state that is part of a federal country. Amazonia's history is marked by contentious claims, counterclaims, and ambiguities regarding tenure and land rights (Campbell 2015). Therefore, establishing the limits of a private property (to develop a REDD project) can be complicated at best, and violent at worst. Although Acre is no longer known for episodes of violence related to land claims (as it was in the recent past), the implementation of REDD projects has revived some land conflicts. Furthermore, as in other Amazonian states, Acre's territory includes conservation units under both federal or state management, indigenous lands managed by the federal government, settlement areas,[3] and public and private lands. By enacting a REDD program that covers the entire state, Acrean authorities began to face the difficulties of trying to reconcile all these different stakeholders under the common goal of reducing deforestation. As one of my interlocutors stated, it was a "tipping point" moment for the SISA program.[4]

"NESTING" PROJECTS INSIDE ACRE'S JURISDICTIONAL PROGRAM

Acre's program of jurisdictional REDD was initially conceived with the aim of targeting international buyers in the voluntary carbon markets. Accordingly, the offsets generated need to be certified by international bodies like the VCS before being traded. The certification process is what assures potential buyers that the offsets were generated following established rules and methodologies of carbon accounting. Certification is thus meant to guarantee that offsets correspond to "real" emissions reductions, and that these reductions have been generated without causing environmental or social harm (i.e., safeguards compliant). The VCS and the CCBA[5] were the bodies responsible for certifying Acre's program and for providing Acrean authorities with the methodological options to integrate the existing private projects in the state's program. This integration is called "nesting" and is mostly intended to harmonize the carbon accounting methodologies and reference levels (Kashwan and Holahan 2014) in order to avoid the double counting of carbon offsets, which would compromise the environmental integrity of both the state program and the private projects. According to the VCS's website, their nesting standard, called Jurisdictional and Nested REDD+ (JNR), "provides robust and transparent accounting and verification approaches for the integration and scaling up of government-led and project-level REDD+ activities."[6]

One important feature of nesting is that it forces authorities to decide whether the projects will be able to trade their credits independently from the state or if the state will centralize the trading and pay private projects their part. This decision is important not just because it regulates potentially lucrative commercial transactions, but also because it has thorny implications for states' sovereignty. During my initial fieldwork it was already apparent that Acre's authorities were anxious that the federal government might impede their own authority to directly trade the state's carbon offsets by arguing that Acre's carbon is a national good (see chapter 4). Once the 2015 Paris Agreement established that developing countries ought to contribute to emissions reductions and that all countries could use emissions reductions from the land and forest sectors, any private projects or subnational programs were potentially turned into direct competitors of national governments. As such, would the offsets generated by private projects inside a subnational jurisdiction such as Acre be traded in a voluntary market or would they be claimed by national authorities to be used as part of its NDCs? In yet another twist with significant implications for sovereignty, might Germany (which has been financing Acre's program since its beginning) actually be able to claim these offsets instead of either Acre or Brazil?

The nesting methodologies created by VCS do not solve any of these conundrums, nor do they mention how to conciliate the competing interests of private project developers, subnational, and national authorities. Instead, VCS's methodologies focus on the technical aspects of carbon accounting and articulating accounting methods, baselines, and reference levels. Ultimately, and despite the conflicts of interest between private agents seeking profits in the market and national authorities complying with international agreements to reduce emissions, VCS works simultaneously for the private projects' developer and Acrean authorities, validating and certifying both the REDD private projects and the state program.

The first time I was in Acre, authorities were working closely with the VCS to create an appropriate nesting standard. The goal, I was told by state officials, was to have "high-quality" offsets, certified by a well-known and reputed international certification body. According to this standard, the existing private projects would have to be registered in the state's registry to be equally certified and avoid double counting of offsets. I asked what would happen if the project developer refused to comply with these rules and did not register his projects in the state registry since his own projects were already certified by VCS and the CCB standards. Márcia, at that time an IMC officer, began by stating that project developers had no reason to refuse the registration because otherwise their projects would be considered illegal by state law.

I also asked what state authorities could do if any of these private projects did not comply with the social safeguards. At that point, things could be more complicated, she explained. Private projects are developed in private properties, so authorities are limited in their action—even when there are squatters in those properties.[7] Furthermore, what according to Brazilian laws could be considered a violation of safeguards might not be seen in the same way by an auditor. Márcia gave concrete examples: since Brazil does not have any legislation on REDD, an auditor can actually check off the box that says that the project is complying with the national law—*because there is no national law*! Another example: the auditor wants to assess the ownership of the property and the project developer presents a promissory sale contract. The auditor will then approve the assertion that there are no conflicting claims over the property, even though such a document has no legal value to demonstrate ownership in Brazil.[8] According to Márcia, it was precisely because of these local specificities—not addressed by the VCS—that authorities were working closely with that institution, in order to create a new standard to be applied to Acre's jurisdictional program. Indeed, this cooperative work between authorities and the VCS to create a new standard was several times mentioned by other state officials as yet further evidence of Acre's pioneering in REDD.

Two years later, though, the nesting process seemed to be more complex and affected by more variables than the mere voluntary compliance of the project developer to have his projects on solid legal ground. Following VCS's rules, Acrean authorities decided that private projects would be able to sell their offsets independently (while the state would retire those credits from their state account). Throughout my entire period in Acre, the registration of these projects never took place, apparently because the state registry was not prepared to start its activity—something that prompted the critiques of some environmental NGOs that considered Acrean authorities to be moving too slowly with this process.

What became clear at that point, though, was the fact that the private projects began to be perceived by local authorities as unwanted competition as well as a liability in terms of public perceptions regarding REDD. This situation was explained to me by Eduardo, who had been very involved in the design of the SISA law and maintained a position in the State Commission for Validation and Monitoring (CEVA) of the law. We had been talking about the four private projects in the state, all developed by the same company:[9]

> The same company! And none of them—this is important to mention—are registered under the SISA. One of them is halfway ready and has land property issues, and social problems regarding safeguards. The [name of the project] has a methodology, I don't know if you read it, it is worth reading, it has a methodology that is extremely fragile. So, what does it say? The landlord has the right to deforest 20 percent, so his avoided deforestation is 20 percent....
>
> ... We had a terrible discussion with the VCS. With the CCB, the CCB gave the greatest certification [to this project] without listening to us! I said, *you shouldn't have done this, you must listen to us, who are here in Acre. And the SISA is for that, it was made with that purpose. You have certified projects with problems, and that is bad for you and it's even worse for us.* We could have developed our own standard, but we wanted to be internationally recognized, because the VCS and the CCB are internationally recognized. So, we have been working with VCS since 2009. We helped them build their jurisdictional standard. And we consider that those certifications [to the private projects] were a terrible mistake.[10]

Throughout our conversation, Eduardo never referred to these projects as market competitors, preferring to focus his critiques on the certification process. However, the American proponent of these projects expressed to me his dissatisfaction regarding the shift in the state authorities' demeanor. If in the past state authorities were keen to meet him every time he traveled to Acre, more recently they avoided him or required him to schedule appointments months in advance. According to him, this change occurred around the Rio+20 Summit,[11] and was all about the fact that his projects and the state's jurisdictional program are competing for the same international funds.[12]

TECHNICAL PROCEDURES, SCIENCE, AND STANDARDIZATION: DENYING LOCALITY

The fact that VCS certifies projects that might not be compliant with state, or even national, law while also providing the certification for the state's program is not seen as problematic by the VCS. When I asked the VCS person responsible for both of these projects and Acre's program certification about these issues, he denied any fragilities in the methodologies used by the projects, considering the accusation "ridiculous." Instead, he pointed out the slowness of authorities in setting the SISA infrastructure and in establishing the necessary registry for private projects. He also denied that the JNR standard had been built in collaboration with Acre's authorities or that the VCS had to, at any point, consider local circumstances and adapt their procedures to create the state's standard. According to him, VCS gave the instructions on how to proceed, and authorities complied. In his view, the only special circumstance of Acre was that the state program would undoubtedly serve as a model for jurisdictional programs across the world. Finally, when I explained that state authorities were upset because the problems with the projects were being used by local activists to attack the state program, he answered dismissively that authorities should have instead chosen a different nesting methodology.[13]

VCS's dismissal of these contentious issues can be interpreted as the expression of an anti-politics attitude through which contentious and political issues are simply disregarded under the banner of bad methodological choices, as if the science behind these carbon accounting methodologies could match the various local circumstances where they are applied and answer to any of the political questions arising from the multilevel policymaking involved in carbon emissions reductions. While these methodologies are very detailed regarding all the technical issues involved in carbon accounting, they are vague or defer to other documents when addressing political questions such as land ownership, law enforcement, conflict resolution, or violations of populations' rights.[14] In sum, VCS's methodologies are able to deflect structural problems inherent to REDD's design by decontextualizing those methods, and it is this decontextualization that gives VCS its political effect (Shore and Wright 2015, 27), that is, its authority as an international certifier. VCS's bureaucratic and complex procedures are, thus, a form of exercising authority (Feldman 2008, 15) and by invoking their expertise and their science-based and standardized methodologies, VCS technicians are able to deflect criticism or even questions related to the implementation of their own rules.

For example, and regarding the specific case of Acre, VCS does not provide any guidance on how grievances expressed by communities inside these private projects might be addressed; nor on how to proceed regarding carbon accounting when this developer does not provide authorities any information about the number of offsets generated by his projects (so they can retire them from the state's account). By invoking their methodologies and technical procedures, VCS officers can ignore all these issues (or refer to them as "ridiculous"), while providing certification to projects and simultaneously continuing to work with Acre's authorities. As long as private projects and state authorities fulfill and check off all the template boxes, VCS can certify them and get paid.

The problems faced by Acrean authorities in their attempts to harmonize private projects with their own program is a clear example of how REDD weaves around itself a complex network of people from the public, private, nonprofit, and business sectors, all of them mediating the local, national, and international levels, according to their different (and at times contradictory) imperatives. And while some of these actors sitting at their desks in the United States or in São Paulo[15] can deflect contentious issues with technical and scientific jargon, authorities in Acre still have to address the criticism and problems generated by the private projects and the SISA law. If, in the case of the private projects, they can claim that there is nothing they can do about it,[16] then regarding the SISA law, those critiques require that authorities reframe the narrative legitimizing their policies and their effects, while still showing international donors that they are committed to reducing deforestation.

Noteworthy is the fact that like the VCS deploys a technical and scientific language to dismiss political issues as well as all the questions related to local circumstances, Acrean authorities themselves bemoan the No-REDD activists for being too ideological, too political, and for refusing to discuss REDD on a technical level. In short, the "anti-politics machine" is deployed at all levels of the REDD network in order to make REDD possible (Ferguson 1994). These multiple efforts conducted by international, national, and local actors to render all REDD aspects technical has, of course, very political effects (Li 2007): it simultaneously precludes any discussions that might challenge REDD's implementation, while also providing increased legitimacy for the intervention of international actors—deemed as those with the necessary technical and scientific expertise. In the end, it is this technical rendering that authorizes and justifies more transnational interventions.

Besides neutralizing its critics, the use of technical and scientific jargon to implement REDD fulfills yet another important goal. By deploying the scientific language of carbon accounting, state authorities are able to include a series of different activities that they consider to be important for Acre's development—some of them already taking place prior to REDD's arrival—even if such activities seemed to be counterintuitive to the purpose of reducing deforestation. In short, the use of technical and scientific language is what allowed state authorities to reinterpret and instrumentalize REDD's template under the SISA law in the pursuit of their own goals—in this case, to pursue their development agenda.

For international donors and REDD advocates, this local reinterpretation and instrumentalization is irrelevant, even if the activities proposed under REDD do not reduce emissions. Independently of the local reasons to take on REDD, what matters for international actors is that, first, its implementation has opened new possibilities for the intervention of transnational actors through the institutional and legal arrangements created by the SISA law; and second, REDD's deployment has granted increased legitimacy to the broad narrative that international actors have consistently sought to sustain and reproduce; that the global problem of climate change can be solved through and in the forests of the Global South. But the local reinterpretation of REDD and its subsequent implementation in different activities is not exempt of problems, including those related to carbon accounting, which requires a more detailed examination of the different components of the jurisdictional program implemented under the SISA law.

Trees, Nuts, Rubber, Fish, and Cows: Acre's Jurisdictional Program

As I described in chapter 4, the SISA law was created in the context of an international program called REM (REDD for Early Movers), funded by Germany and intended to support REDD's implementation, with the perspective of a potential REDD international market starting in 2020. The program involves several specific lines of intervention.

The first one is the expansion of sustainable forest management into areas that were previously removed from any type of timber extraction, namely extractive reserves.[17] The rationale informing this policy is that, since deforestation is unavoidable, it is better to carry it out legally and sustainably than to leave it to the mercy of illegitimate actors. The government also seeks to promote the extraction of nontimber goods (such as rubber and the Brazilian nut), claiming that these measures will ensure that these communities have enough resources, preventing them from cutting down trees. The government also supports the settlement of families with agricultural training and provides them with seedlings for fruit trees to promote reforestation. Yet another important component of the state's program is the promotion of fish farming, through a factory that produces and sells fingerlings and food, buys back grown fish, and processes and sells them on the Brazilian market. State officials argue that fish farming is not only a very efficient way of producing protein, but that it also prevents further deforestation by offering a far less space-needy alternative to cattle ranching. Following the same logic, authorities are also supporting more efficient forms of cattle ranching, improving and rotating pastures, and supplementing cattle nutrition. A final major component of the program is the implementation of management plans on indigenous lands as a way of rewarding indigenous populations for their role as forest conservationists. I will provide a more detailed description of the program's different components while demonstrating how the local reinterpretation of the REDD template can ultimately contradict the stated purposes of this conservation mechanism.

INGRAINED CULTURE AND UNFULFILLED PROMISES

The state of Acre has a long history of timber extraction, having started what is called sustainable forest management around 1995—well before REDD was part of climate negotiations. Although seemingly contradictory (since REDD is supposed to leave trees standing), the inclusion of forest management in REDD has recently been accepted within the climate epistemic community (Long 2013). Indeed, during fieldwork I asked several people about the integration of sustainable forest management in REDD, which seemed contradictory to me. Eduardo, Danilo, and Vagner stated vehemently that it was not a contradiction at all, it just depended on the right carbon accounting methodology. I later found the same explanations in Mozambique. The scientific argument for including logging in REDD is that the removal of older trees promotes the growth of new ones, which increases forests' carbon sinks. However, this is a debatable topic within the scientific community: if young trees remove carbon from the atmosphere more quickly than old trees, their carbon reservoirs are smaller; old trees capture carbon at decreasing rates but maintain larger carbon storages. The carbon balance between young and old trees is, therefore, far from scientifically settled (McSweeney 2014; University of Minnesota 2016).

According to Vagner, one of the leaders of the cooperative formed with the purpose of expanding this logging policy,

> if you have an activity of forest management in which in each hectare you take up to one, one and a half trees, and it is scientifically proven that when you cut a tree, in that clearing the remaining trees will have a faster, better, and healthier growth. Then scientifically, forest management is beneficial to the forest. I have no doubts about that. It is proven.[18]

Thus, following this logic, state authorities created a cooperative that would aggregate all the timber collected from the state, seek its international certification as sustainable timber, and sell it in international markets. The expansion of logging into the extractive reserves would ensure the consent of each family to allow the removal of the most valuable trees from their settlement in exchange for the money acquired through the sale of the timber. Elements of this cooperative promoted meetings with families in the reserves during which they emphasized the benefits of sustainable forest management, arguing that the income provided by logging could be an important element for these families' livelihoods.

However, some communities inside the extractive reserves do not share this positive and scientifically proven view and have resisted the expansion of sustainable forest management into the reserves. The most outspoken critic of this policy is a member of the rural workers' union in Xapuri, Daniela, who participated in the violent struggles of rubber tappers against the deforestation promoted by cattle ranchers during the 1970s and 1980s (see boxes 1 and 5). In her view, this policy is not only an offense to the legacy of Chico Mendes, but also a foolish endeavor that fails to address the economic problems of poor families. The forests will be destroyed, and thirty years will not be enough for the forest to recover from the "management," she says.

Moreover, the money paid to families by the cooperative is so minimal as to render it insulting, and an activity that is supposed to be communitarian is actually being carried out by business companies.

Box 5

Operation Amazonia: Developing the forest

The launching of Operation Amazonia, in 1967, responded to two main sets of concerns by the military authorities: state control and development. As such, the plan extended the presence of the military into its most remote areas of the forest, promoting the colonization of a land portrayed as void of people. The military presented the internal migration of people from other parts of the country to Amazonia as the solution for the problem of poor families dispossessed by processes of fast industrialization and agriculture expansion. The operation, thus, launched a whole new economic paradigm in the region: instead of a familiar model of subsistence based on forest extractivism, the military intended to clear the forest, opening it to cattle ranching and industrial agriculture (Hecht and Cockburn 1990).

The modernization model envisioned by the military had immediate consequences in Amazonia's physical and human landscape. Rubber settlements were taken over by the government and delivered to southern entrepreneurs willing to clear the forest and convert these lands into new farming and cattle ranches. Previously densely forested areas gave way to naked fields punctuated by solitary nut trees—still considered profitable—and Nellore cows. Fire and bulldozers followed the cut of the most valuable trees, and those who tried to resist these methods were persuaded at gunpoint.

In this context, social tensions and increased violence opposed landholders (supported by the military), and rubber tappers and other "forest peasants," who sought to resist and revert a development model that clearly meant the end of their livelihoods in the forest. Acre became the center of many of these fights, with the creation of the first rural unions (Paula 2016). Unions became important in organizing the fight and connecting rubber tappers to outside sources of power (Almeida 2002, 185)—especially at a time when the violence of the military dictatorship had reached unprecedented levels, with arbitrary killings, massacres of indigenous communities, constant threats against activists and community leaders, and evictions of entire families.

Figure 8 Cattle in a deforested area near Xapuri. The only trees standing are Brazilian nut trees, as there is a ban on cutting such trees. However, these die after some time because their survival depends on the existence of other trees around them. Photo taken by the author. Xapuri, 2015.

I had the opportunity to travel with Daniela to one of the extractive reserves close to Xapuri to talk to some residents about their experience with sustainable forest management. After an hour of many bumps, potholes, and dubious wooden bridges, we met at the house of one rubber tapper, who was accompanied by two others. The three men (in their forties and fifties) were born in the reserve and still extracted rubber, even though this activity was no longer enough for them to make a living. Sustainable forest management had been introduced to them by people from the cooperative during a whole-day session about the benefits of allowing that practice in their settlements. Some people got excited about the money that was promised. These three men, although initially skeptical, were convinced—mostly because they needed the money. They were promised a certain price for the timber, a new road (needed to transport the logs from the reserve), and a small dam to save water. "They even promised us a woman each!" one of the men added, making everybody laugh. But none of that came to fruition, they concluded.

After the first round of logging, the three men (among others) dropped out of the cooperative, feeling duped over the money they had been promised but ultimately never received. In the end, they said, the timber cooperative was just like the rubber cooperative the government had launched years ago: it went bankrupt, and nobody got their money back.

For Vagner, the resistance of reserves' inhabitants to forest management is not only a sign of ignorance and of cultural backwardness, but also a political reflex of simply opposing the current government. It is, according to him, an ideological matter that gets in the way of science:

> So that logic remained, that cultural logic of them, let's say, the idea that cutting timber equals destruction. That idea got ingrained popularly, culturally, that taboo remained and was not torn down. And that reflects in the [lack] of acceptance towards sustainable forest management. There are violent critiques from people who lack technical knowledge, vote ignorantly, and sometimes are moved by political reasons. People ignore the technical part of it and adopt a political stance.[19]

Vagner's words provided a rich example of how state authorities deploy a technical and scientific discourse to legitimize their actions—even when the science is not settled. For Vagner, the scientific rationality of state officials' policy was thus being opposed, not by a competing scientific argument, but by sheer ignorance and cultural backwardness.

In the end, and independently of the positive or negative impacts of sustainable forest management for both the forest and the goal of reducing emissions, this policy presents two problems. The first one is the competition from illegal wood. The market is saturated with cheap illegal timber, and the cooperative cannot compete with the prices of unsustainable logging, which means that the market itself will necessarily push agents toward less sustainable practices of logging. The second, and perhaps more important, is that by expanding logging into the reserves, authorities are also opening these territories to other actors who might exploit it unsustainably. That is, while claiming to be protecting forests (and thus reducing emissions), state authorities are actually promoting their deforestation by enabling logging where it was always banned. In the meantime, Daniela (and a few others) continue to fight against authorities' plans of expanding sustainable forest management into the extractive reserves, even if that means going against other union officials in Xapuri and other rural unions—all of which have, according to Daniela, been "bought" by the government.

THE IMPOSSIBILITY OF MAKING A LIVING OUT OF RUBBER

As with timber, the state government has also been pursuing a policy promoting a pro-market approach to nontimber products in an effort to diversify families' sources of income. This policy has been mostly focused on rubber, açaí, and Brazilian nuts, and includes training, technical assistance, and the support of a cooperative to help reach markets. However, Daniela and others also criticize state authorities on this endeavor, for lack of consistency. While families inside the reserves are criminally prosecuted and expelled from the reserves for keeping cattle, Daniela points out that, unfortunately, cattle ranching is the only activity that can ensure these families a decent livelihood. In an interview for a local blog, Daniela argued that cattle ranching was, and continues to be, consistently promoted by the federal government, and state authorities have not pursued a cohesive policy that can actually add value to nontimber products and be an alternative to cattle. She went into specifics:

> I dare former rubber tappers who are now advocating for extractivism, to sell a kilo of rubber to the condom factory in Xapuri, for R$4.10. The average amount of rubber produced in 30 days is 100 kilos, which makes R$410.00. Is that money? It's less than the minimum wage. (Machado 2008)[20]

Families have no other option than resorting to cattle ranching, she concluded.

The impossibility of making a living out of rubber tapping was a topic of discussion during my conversation with the rubber tappers. While rubber had once been Acre's main export, rendering this small Amazonian state the status of world producer, planted trees and synthetic rubber turned the traditional tapping process almost useless (see box 6). The factory, built by the government in 2008 in Xapuri, to ensure a market for this product, only pays R$8 per kilo—it has been the same price for many years, they complained—and at that time, they were also concerned about rumors that the government was going to close the factory for lack of economic viability. More recently, state efforts to promote the cultivation of rubber trees only made things worse for those tapping wild trees. Cultivated rubber is more profitable, they explained. In one day, a cultivated field produces the same amount that rubber tappers produce in one week. To corroborate what the rubber tappers were saying, our driver added the example of a man he knew who was able to sell 120 containers of 50 liters each, each week. His six workers were paid according to their productivity and made more money than if they were tapping their own rubber settlements.

Box 6

Rubber production in Acre

To understand the constitution of Acre as a distinct region inside the Amazon, it is necessary to acknowledge the importance of rubber production, how it dictated the economic cycles of the region, and its interdependency vis-à-vis international centers of power. The extraction and production of rubber in the Amazon were based on a primitive model of labor that changed very little throughout history. Rubber tappers were mostly impoverished migrants from Northeastern Brazil, sponsored by rubber merchants from Manaus and Belém, who would support all the costs of their dislocation in exchange for their later payment through work and rubber. Rubber production depended upon and reproduced a multigenerational chain of indebtedness: these merchants—indebted to English and American buyers—supplied the landholders with all the commodities provided to rubber tappers in exchange of all the rubber collected. And each landholder kept his workers (rubber tappers) indebted using the rubber collected by them as payment (Almeida et al. 2002; Iglesias 2010).

The expansion of rubber production and the transformation of Amazonia into the biggest world supplier would not have been possible without the funds from companies from Liverpool, Hamburg, and New York, who monopolized rubber production and trading (Neto 1979, 88). However, while Amazonian rubber supplied important thriving industries, the structure of this economy only increased Brazil's dependency and subordination to the interests of Americans and Europeans (Neto 1979, 73). The first rubber boom occurred between 1880 and 1912. British investors dominated the whole business from 1860 to 1900, when Americans began competing for rubber trading. In 1900, Amazonia exported almost twenty-seven thousand tons of rubber, reaching its peak in 1912, with 43,370 tons. After that, production started to decline due to greater competition from Asia. With the beginning of World War II and the control of the

> Asian plantations by the Japanese, Amazonian rubber went through another boom. American authorities turned to the Amazon, ensuring a quick upgrade in rubber production, and exclusive access to it at low prices (Garfield 2013, 51).
>
> When the war ended, the demand for rubber declined and the Asian plantations became, once again, available to Western markets. Consequently, the rubber economy in the Amazon crashed. These cycles of boom and bust contributed to a significant change in how Brazilian authorities began to understand the region. After long periods of neglect by central authorities—mostly for lack of a consistent idea about how to manage the region—federal authorities could henceforth see its economic potential.

Despite the efforts of state authorities to maintain rubber tapping as a source of sustainable livelihood, the failure in maintaining the factory opened became a clear demonstration of how local initiatives are never independent of international commodity chains, and therefore can hardly be successful when implemented disconnected from those chains. Natural rubber became a niche product, whose demand is not enough to maintain the extractive activity. Noteworthy also was the fact that, even though state authorities claimed that the monetary benefit paid by the state in exchange for the rubber tapped inside the reserves was part of the SISA law, such benefit had in fact been created years before REDD began to be discussed in the international arena. In the end, it is apparent that while traditional rubber tapping is no longer a viable way of making a living, cattle ranching constitutes the biggest competitor vis-à-vis any other activities developed inside and outside the reserves—and also one of the greatest drivers of deforestation.

FISH FOR COWS

During my conversations with different state officials, I was encouraged to visit the "fish factory" recently built in the state. Introduced to me as one of the biggest innovations in Acre's program, fish farming has, however, been part of development initiatives since at least the 1940s, as a technology with the potential to improve nutrition, generate income, and diversify livelihoods (Crewe and Harrison 1998, 8). In the 1980s, support for fish farming declined due to its lack of success in improving livelihoods (8). Yet, and despite its controversial history, fish farming in Acre was repeatedly highlighted by local authorities as their showcase initiative in fighting deforestation:

> Fish farming has a protein yield per hectare one hundred times higher than cattle. The average head of cattle per hectare is one. You need an area of 10 thousand square meters for one cow. Whereas in one hectare, ten thousand square meters of water you can produce a lot of fish![21]

Roberto then proceeded to explain how the factory, while a private endeavor, was built with funds from the National Development Bank (BNDES), the state government, and a private investment fund, benefiting from state policies. Despite all the hope for this entrepreneurship, Roberto was well aware of the challenges:

> There are good signs, there is clarity about what should be done, but in practice, the difficulties are big because of the amount of variables. Nowadays, because of globalization, you're part of a system—a system in which you cannot control all the variables. So you might be pushing for a business, but if the demand is different . . . the Chinese started eating more meat. You can turn yourself inside out to sell fish, but it won't work, get it? They want meat. And if they want meat, your deforestation is going to increase. There is no way around it![22]

Roberto's statement clearly points to the same problem affecting the aforementioned policies to promote rubber tapping and sustainable logging: demand for all of these commodities (be they timber, rubber, fish, or beef) depends on international markets, regardless of local efforts to promote or discourage their production. Thus, no matter how much Acre's government promotes fish farming, if international markets demand beef, the fish factory will fail, and deforestation will increase.

Roberto's words were also significant for highlighting what is often obscured in environmental discourses: the drivers of deforestation are complex and are only partly local, involving different economic sectors. In the case of Acre, even if promoting fish as a protein replacement could have effects in the demand for beef at the local level and even national levels, there is still an international demand for beef, soy, corn, sugar, and timber that pressures the entire Amazonian region. From this perspective, a mechanism like REDD, that is implemented at the local level and never addresses some of these international drivers, can only be doomed to fail in its stated environmental goals.

SUSTAINABLE COWS?

Perhaps because of the recognition that cattle ranching is an overwhelming force, hard to compete against, the government has also tried to develop techniques to enhance the yields of head of cattle per hectare. Cattle ranching has pervaded Acre's economy, not just on a large scale—against which rubber tappers mobilized—but also on a smaller scale, inside the reserves, as a means to ensure a livelihood. The increase in cattle ranching in the reserves should not be seen, however, as a symmetrical counterpoint to the shrinking role of rubber tapping in the families' economy; rubber and cattle are distinct commodities serving distinct purposes, and while the former was converted into daily consumption goods, the latter fulfilled long-term saving purposes (Pantoja, Costa, and Postigo 2009, 121). Ironically, it was the success of the extractive reserves in providing families with economic stability that enabled the constitution of some savings, and, for lack of available alternatives (such as bank accounts), cattle served that purpose (121).

The increase in cattle ranching has been so pervasive since the early 1990s,[23] that it is now possible to identify throughout the state a "cattle culture" followed by a fascination with cowboys—especially among the younger generations (Hoelle 2015). These new generations have no interest in pursuing rubber tapping—or any other activity carried out by their parents, for that matter—and see in cattle ranching the fastest way to obtain coveted material goods, and even to get rich. Cattle ranching is associated with prosperity and with a modern lifestyle that is opposed to the hardships of forest-based and agricultural livelihoods (Hoelle 2015, 5).

For those opposing the state's policy, like Daniela, cattle stocking is necessarily indexed to the history of the state, marked by the violent and bloody fights against cattle ranchers (see box 5). Inside the reserves, the activity is still censurable, she explained, but understandable at a smaller scale, given the lack of economic options. For different reasons, Daniela's opposition to state initiatives targeting more sustainable yields in cattle ranching matched some concerns expressed by environmentalists. Among environmentalists in general, there is a long historical tradition (connected to colonial anxieties regarding overgrazing) of seeing cattle stocking as a threat to the environment (Netting 1977) because cattle herders tend to deforest to increase grazing fields. When I sat down with Carlos, a member of an international NGO that works with state authorities in the SISA implementation, he told me that the use of funds from Germany to promote cattle ranching—even if in more efficient ways—was regrettable. For Carlos, this not only demonstrates lack of coherence in state policies, but also reinforces his fears that, given this policy's tipping point, the government is about to give up on the sustainability paradigm in favor of cattle.

Cattle ranching constitutes a double bind for state authorities. On one hand, it embodies an ethical position rooted in the history of the violent struggles led by rubber tappers against cattle ranchers. Under this position—subscribed by both environmental activists and those on the left side of the political spectrum—cattle ranching is inherently negative. But on the other hand, cattle ranching is an important economic activity that provides significant income, and that is increasingly perceived as the means to achieve better livelihoods—especially among the younger generations. The claim to be seeking better ratios between heads of cattle and grazing areas represents the attempt of conciliating these two opposing views within a context in which both logging and cattle ranching are considered by authorities to be inevitable.

WELL-BEHAVED INDIANS: REDD IN INDIGENOUS LANDS

One of the basic rules of REDD is the principle of additionality, which means that there must be deforestation already happening to make the case that without the implementation of REDD, deforestation will continue or increase.[24] The reduction of emissions is thus generated from this difference between what was supposed to happen (had REDD not been implemented) and a new reality where REDD is doing its work of reducing deforestation. Given the fact that indigenous lands in Brazil are the ones where deforestation rates are the lowest, the implementation of REDD in those territories seems nonsensical. Nevertheless, Acre's program includes indigenous lands, and state authorities have actively sought to engage indigenous peoples, setting an indigenous task force to take on REDD.

The involvement of indigenous peoples in REDD did not begin in Acre, and in fact, a US-based NGO has been very active in the development of REDD and other payments for ecosystem services among several indigenous communities in other Amazonian states. I met a director from this NGO, Bruno, who is based in California but traveled to Acre to organize an event promoting Yawanawá craftworks.[25] Bruno is Brazilian, and despite being connected to several US environmental NGOs for many years, his work has been focused on indigenous people in the Amazonian region, notably Acre, where he has several friends. We talked over Skype after his return to California, and I asked him about this seeming contradiction of REDD in indigenous lands.

> I am a bit critical of REDD's classical model. I think we need to revisit REDD's classic model in regards to the issue of additionality. Because the additionality [rule] leaves aside many areas, if not all the indigenous territories located in remote areas, because there is no deforestation there. So, I think this is the Achilles' heel in REDD's methodology. From the technical point of view, strictly technical, one understands. But from a humane point of view, from a socio-environmental point of view, from an integral management point of view, and from a long-term perspective, you don't know where the new deforestation drivers will come from. Honestly, I think it is a myopic mechanism.[26]

In a Western imaginary, indigenous people around the world have historically fulfilled the role of natural conservationists, of being closer to nature (or part of a pristine nature), and therefore, of being naturally ecological (see Krech III 1999, Garfield 2004, Shah 2010, Cunha 2012). Notwithstanding the problems inherent to these misconceptions essentializing indigenous peoples as a different category of people (separated from culture and from a history of globalization and capitalism), NGOs, donors, and development agencies have—sometimes unwittingly—perpetuated this logic. Bruno considers REDD myopic precisely because it does not reward the *natural conservationism* of indigenous peoples.

The gradual transformation of REDD from a market mechanism (what Bruno called the classical REDD) where additionality counts, into a payment-per-results system in jurisdictional areas enables the rewarding of indigenous people for their role in the conservation of forests—because, in the end, what counts is the amount of avoided emissions in an entire area, of which indigenous lands are just a part. The SISA law contemplates the rewarding of indigenous peoples in accordance with their Territorial and Environmental Management Plans (PGTA), also known as "Life Plans."[27] As such, authorities have funded programs to certify Indians as agroforestry agents, to help them implement their PGTAs, and to promote training.[28]

While in Acre, I had the opportunity to talk several times with a government aide for indigenous affairs. Very knowledgeable of indigenous issues and history, Lucas provided me with a brief chronology of the indigenous movements and of their organization to work toward having their rights recognized. This movement, which involved academics (mostly anthropologists) and NGOs, started in the 1990s and generated a series of public policies in health and education. Gradually, and in relation to the state's economic zoning, these indigenous movements (in all their diversity) evolved into the preparation of their Life Plans. The beginning of the REDD conversation emerged in the context of the consultations for these plans. Lucas explained:

> So, for the people involved, that made sense, and if there was a conversation [about REDD] it was only possible to the extent that it made sense in relation to all the things that were being discussed: ethno-zoning, management plans, agro-forestry agents.

To Lucas, REDD or the SISA law do not change Indians' practices:

> That story that people talk around here, that they [Indians] will have to reduce deforestation, that there will be very concrete goals, that you will have to measure a bunch of things, all that stuff, I don't think that's the way. On the other hand, there is no way Indians can stay out of this policy. As I said, the indigenous program does not take into account that quantification of carbon, those criteria of good behavior, of committing to not deforest, you know? It's not that type of thing.[29]

According to him, what the SISA law has been doing is simply the continuation of public policies that had been previously agreed upon with the indigenous peoples. In fact, he stated, the SISA law should not be seen as something "completely new" but rather as a "catalyst for strategies"—in the case of the Indians, "for accessing funds." The same view was expressed to me when I visited the Pro-Indian Commission, an NGO founded in 1979 to support indigenous peoples and advocate for their rights. In my meeting with two staffers (one Indian man, and one white woman) I realized that none of them had an idea of what REDD is, how it works, and how carbon is related to it. For them, the SISA law is a mechanism to deepen the relationships with state authorities and advance indigenous peoples' claims. Accordingly, they demanded that agroforestry agents be recognized as a professional class, with a contract and a regular income (instead of being paid according to projects' funds or Germany's donations) because it is their work that prevents fires and detects illegal logging, thus preventing deforestation.[30]

These perspectives on the SISA law are certainly not shared by all indigenous groups.[31] For some indigenous leaders like Ninawa, the leader of Acre's Huni Kuin People Federation,[32] REDD is a deplorable way of commodifying nature that perpetuates a long history of depriving indigenous peoples of their rights. As a community leader, Ninawa travels around the world advocating for indigenous rights and fighting against schemes like REDD which, in his opinion, is yet another instrument used to take away land from indigenous peoples. For Ninawa, REDD and similar schemes involving the commoditization of nature are mere deceits sponsored by the big polluting industries around the world, who claim to be concerned about the environment, while not doing anything about it.

Alternatively, for José, a Yawanawá leader, it is an opportunity to right those historical wrongs:

> The indigenous people already provide a very relevant service to humanity. And what we want . . . what we want is that indigenous people are rewarded for that. It's something that we do *naturally*. And if we can get money, be rewarded to continue developing our activities, that will only strengthen our work.[33]

José was very strategic in the deployment of Western stereotypes about Indians, as a political instrument to make claims—what Carneiro da Cunha called "culture" within quotes (Cunha 2009; see also Pantoja 2014). Here, he referred to the supposed closeness of Indians to *nature* (see emphasis in quote) expressing a new form of environmental subjectivity arising in reaction to the new rules regulating nature—what Agrawal (2005) called environmentality. Indeed, his discourse reproduced some tropes found in Western environmental discourses advocating for payments for ecosystem services. Yet, it would be naive to interpret his words as evidence that Indians are merely being subjugated in this process. In contrast, this young, but very well-traveled leader, with a deep knowledge of transnational logics of aid, development, and policymaking reveals how these communities are also able to locate new spaces for political maneuver and exercise political agency within the terms of this discourse. Back in 2013, José spoke enthusiastically of his plans to develop a REDD project in Yawanawá territory, of how he had already committed American buyers for the carbon offsets, and of how this project was just one among several other collaborations that the Yawanawá have developed with American and European corporations throughout the years. By 2015, José's views reflected his increasing concern with deploying REDD to counter state authority:

> We, the Yawanawá, have worked our project of payment for environmental services in connection to the Yawanawá life plan, thinking that the government had the platform [the registry for private projects]. There's no platform. So, everything that they have been saying around, to the international community, is a lie! To me, it's all a lie. Because during our last conversation [with the state governor], I told him that the government had to give us the carbon credits that belong to us, so we can negotiate directly with the market. And then, that changed completely.
>
> [Me] *So, the idea of having the government giving you the benefits from those credits through public policies is not worth it for you?*
>
> To us, that doesn't work. We want to do it ourselves. Ourselves!
>
> [Me] *Why?*
>
> Because the government's program is charity, so the indigenous people depend more and more on it. We don't want to depend on the government for all of our lives. We want to be independent from the government....
>
> Everything is very clear. There is a study done by MacKenzie[34] and a bunch of lawyers, to understand whether indigenous peoples had the right to carbon or not. And they published that they do. That was presented at FUNAI,[35] in Brasília, at the Itamaraty.[36]

From the first time I met José to the last time we were together, he had been made aware—in a frustrating way to him and his people—that one of the most politically contentious issues of jurisdictional programs is ownership. While José was counting on the possibility of trading offsets from his people's land—a possibility that was hardly on the table given REDD's additionality rule—state authorities have been using indigenous lands to balance off the high deforestation rates occurring in other parts of the state. In a jurisdictional program, there is no distinction between land (and carbon) ownership; the only thing that matters is the overall amount of avoided deforestation in the entire jurisdiction. If, for José, the state is stealing Yawanawá's carbon, for state authorities, the Yawanawá and other indigenous peoples are compensating for the deforestation in the rest of the state, keeping the overall rate below the necessary threshold.[37] These contradictory claims are addressed by state authorities through the deployment of the technical and scientific jargon of carbon accounting—such as the one used by the VCS—and by including some indigenous leaders in the policy discussions for the SISA implementation. But the political contentiousness of issues such as carbon ownership, carbon rights, and political autonomy remains unsolved. As such, José was faced with the fact that, despite his conversations with international NGOs and donors, the state rules for the jurisdictional program (and now federal laws, too) preclude him from selling carbon offsets.

Ultimately, what this situation demonstrates is how the REDD template could be yet again reinterpreted and appropriated in the pursuit of very different goals. In the same way that Acre's authorities took up REDD as a means to access additional international funds, indigenous leaders like José saw in REDD an opportunity to expand his group's political autonomy. As such, he appropriated the mechanism under his own terms and negotiated its implementation in Yawanawá lands with international NGOs and donors—independently (and even against) the interests of state authorities. The capacity of indigenous communities to leverage their own political agenda with international actors, and independently of Brazilian institutions and authorities, has always been a source of anxiety for these institutions and authorities (especially the military). Even if these anxieties are not directly related to issues of national sovereignty, they are a demonstration of how Indians' political agency can disturb common stereotypes about indigenous people (see Turner 2002; Cunha 2012). For international actors, Indians' agency constitutes an important vehicle for them to expand their activities and presence, while also seeking to decentralize processes of decision-making. By working directly with indigenous communities such as the Yawanawá, international NGOs (such as the one where Bruno works) opened an important space for transnational governance.

WE HAVE BEEN TALKING ABOUT THE SAME THING FOR THE LAST TWENTY YEARS: SUSTAINABLE DEVELOPMENT[38]

During an event in Washington, D.C., in October 2016, promoted by Forest Trends Ecosystem Marketplace, I had the opportunity to chat with a member of that organization about how much REDD had changed since its initial definition. He agreed, but also continued to refer to REDD as a market to trade carbon offsets. Then I described Acre's program, explaining that for state authorities, their fish factory is part of their REDD program. I could see in his face how surprised and intrigued he was.[39] It was not the first time I had experienced this disconnect between how REDD is talked about in the World Bank and environmental NGOs' corridors, or among market agents, and what is actually going on in places like Acre and Mozambique. REDD has indeed taken multiple forms and met (or failed to meet) very different expectations to the point of making it possible to implement REDD without people knowing what REDD is or implementing it in such ways that those more attuned to the market logic of the mechanism do not recognize it. Setting aside the people involved in the SISA law and those who fight against it, most people in Acre have no idea what REDD is or that the government is trying to reduce deforestation through a set of policies that reward supposedly sustainable forms of production. Fish farmers do not know that their activity is being subsidized by German funds, rubber tappers selling their rubber to the condom factory do not know that the price they are paid is now subsidized as part of these policies, and residents of the extractive reserves—although very aware of the government's plans to log their settlements—are not that knowledgeable about the link between what is called sustainable forest management and carbon emissions. This lack of public knowledge about the SISA law was in fact criticized by the KfW, who forced Acrean authorities to be more proactive.[40]

All these policies included in the SISA law are about development (be it sustainable or not), as Márcia made very clear during one of our conversations. As such, for Acre's authorities, REDD opened the possibility of accessing additional international funds—much needed in one of the poorest states in Brazil, often neglected by federal authorities—so they could continue to implement the policies deemed more appropriate for the region. That means taking advantage of the existing forest resources (which have been part of the state's economic model for many years, although in different formats and intensity), exploring the potential of some products in international markets (like Amazonian fish or the super fruit açaí), and still catering to one of the biggest driving forces in Brazil's economy, cattle ranching. Unsurprisingly, the activities included in the SISA law that are supposed to provide alternative sources of income are exactly the same ones offered as benefits in the private REDD projects, and the same that have been tried for decades in the Amazonian region, promoted by USAID and other development agencies and NGOs.

The fact that REDD funds are being used to maintain (or expand) activities that were already being developed or that have already been tried (unsuccessfully) in the past raises important questions regarding carbon accounting and the additionality rule. International donors are certainly aware of this potential mathematical flaw. However, the different elements of the SISA law were never contested or their environmental integrity questioned. For donors such as Germany, Norway, or the UK, the implementation of the SISA law responds to more important imperatives than the mere creation of the conditions for a future REDD market. Indeed, the local reinterpretation of REDD (applying it to agriculture, fish farming, or even the management of Indian lands) provided these donors both the possibility of broadening the scope of their intervention and the legitimacy to claim that they are doing something to tackle climate change. This shows how local politics feeds back—and intersects across—the national and international levels of the policy world and how this feedback is also a constitutive part of transnational governance.

Unsurprisingly, the REDD process in Mozambique evolved in very similar ways, moving from a strict market perspective into a jurisdictional program involving multiple sectors and including activities already tried in the past. I will now turn to Mozambique to explore these similarities (and differences), in order to trace some conclusions about their implications to transnational governance.

Landscape Approach: Carrying Yet Another Fantasy

When I returned to Mozambique in 2016, after initial fieldwork in 2014, the country had been through elections and a major transformation in its ministerial organization. The Ministry of Agriculture was no longer involved in the REDD process, and the directorate that previously concentrated all the "forest people" was incorporated into the new Ministry of the Environment, responsible for native forests. This latter ministry had been reestablished as the Ministry of Land, the Environment, and Rural Development (MITADER). Prior to this ministerial reform, the environment ministry held a mere coordination role between other ministries. The new nomenclature of the ministry was significant, as it implied a clear connection between the environment and rural development. This connection was particularly telling given the latest developments of REDD in the country.

The REDD process was no longer split between ministries, fueling institutional rifts, but was instead coordinated by the REDD Technical Unit (UT-REDD), an "elite cabinet"[41] operating under the wing of the new minister. This new minister was a strong figure, a young leader with political ambitions, sometimes referred to as "the tiger," given his connections to the Malaysian financial world—indeed, he spent part of his life abroad and left the board of one of Mozambique's main banks to take on the new role. In this new institutional landscape, the environment seemed to have a more prominent role, and given the newly created UT-REDD, I anticipated that the assertion that REDD does not exist yet, which I had heard two years before, would no longer hold.

The first time I visited the UT-REDD offices, I met Luís, who updated me on the most recent changes on REDD. I was already familiar with the new vocabulary used by the World Bank in the numerous documents about Mozambique I had downloaded from the internet. "Landscape approach" was the new catchphrase, and I was puzzled by how this expression had taken over REDD. More to the point, I was curious about what this shift in terminology was supposed to signal on the ground. Luís explained that the idea of a landscape approach had started to be discussed locally the previous year. The purpose was to conceive of a series of activities to include in REDD that, in the end and altogether, could reduce deforestation. These activities covered conservation agriculture, new and renewable energies, reforestation, biodiversity conservation, and community forest management. However, Luís alerted me that they avoided talking about REDD, especially carbon, when they tried to bring and implement some of these activities in the communities:

> If I go to a local community, I don't want to talk about carbon. That's not the way of explaining to people the importance of REDD. Because people might

get even more confused. No, you have to give concrete examples. So you tell people no, that, instead of doing that (itinerant agriculture), they should do conservation agriculture. And people understand that their production increases. Now, the goal of reducing emissions, *that stays with us, right?* You don't go to a peasant and say, you are [reducing carbon emissions if you do conservation agriculture]—he will never understand.[42]

The purposeful avoidance of the terms REDD and carbon (confirmed by the World Bank team leader and other members of the UT-REDD) was also explained as necessary to avoid raising expectations about any eventual economic benefits to the communities that proved successful in reducing deforestation. In his statements, Manuel, the UT-REDD leader, underscored some of Luís's previous assertions:

If you fail to reduce emissions, you get nothing [meaning, you do not get paid]. Therefore, we avoid talking about benefits. Even knowing that there is a benefit. We avoid saying that there is an economic benefit. We say that there is an economic benefit for him [the peasant] because he increased his productivity, the environmental issues, all of that.[43]

What Manuel meant was that instead of telling peasants that they can get paid for the carbon offsets they might generate through their agriculture practices, people from the UT-REDD convince peasants to shift their cultivation methods by appealing to the higher agricultural yields they supposedly can achieve with these methods.

The World Bank had signed a letter of intent with the Mozambican government, Arnaldo (the World Bank team leader) told me, declaring the bank's willingness to pay around US$5 per ton of avoided carbon emissions through reduced deforestation—if there is indeed any reduced deforestation. To him, that letter and the future agreement[44] to buy emissions reductions are useless, except if taken as an incentive for the government to implement adequate policies to reduce deforestation, even though the total agreed amounts to US$50 million.[45] Regarding the very low price per ton, he explained that the bank uses set prices, "because there is no market anymore. Basically, what you have is the donor's willingness to pay."[46] Therefore, in the absence of a market in which the laws of supply and demand establish the price of carbon, the World Bank sets such prices in accordance with what international donors are willing to donate to reduce deforestation.

It was clear from all the conversations I had with members of the UT-REDD and from the World Bank that nobody was interested in REDD anymore—at least not in its original form of a market mechanism based on the sole activity of avoiding deforestation to generate carbon offsets to be traded at an international scale. The focus had shifted to something rather different—the landscape approach—in which REDD could incorporate many different activities. For Arnaldo, that change was part of a natural evolution:

> I think this paradigm shift to the jurisdictional makes sense. But it is not easily understood by countries. I work in other countries, Ethiopia, Congo, and that shift was difficult. It was a difficult shift. That's what's fascinating about REDD: the way it has evolved internationally. Because I remember, several talks we had here with MICOA that went like this: *the World Bank arrived*—and they would go desperate—*here comes the World Bank with a new idea!* Because we were the carriers of those fantasies.
>
> [Me] *And you still are, right?*
>
> And we still are, but now it's the landscape.[47]

The international evolution that Arnaldo was referring to concerns the gradual shift of REDD from an exclusive project-based market approach into a jurisdictional one in which countries are compensated for all activities deemed able to reduce carbon emissions from the forest sector.

Arnaldo had no issues with admitting that the *landscape fantasy* had been brought to Mozambique by the bank even though I had previously heard some members of the UT-REDD claiming that it was the Mozambican government who had pushed for a more holistic and developmental approach to the rural areas of the country. According to this version, the new minister asked for approaches that could bring development into certain parts of the country, and the World Bank responded to this request suggesting "the landscape." As in Acre, the landscape approach—that is, a jurisdictional program—included many different elements, all of which are supposed to reduce deforestation. I will describe each one of them, underscoring their implications to transnational governance.

Forum in Zambézia: From REDD to Rural Development

Since the beginning of the REDD process in Mozambique, the World Bank's Forest Carbon Partnership Facility (FCPF) had donated a total of US$8.6 million to prepare the country for REDD. According to the information provided by the bank's website,[48] those funds were used to support the creation of the UT-REDD, the preparation of the REDD national strategy, and the public consultations to draft the jurisdictional programs. The allocation of these funds had nothing to do with the initial REDD law approved in 2013 (see chapter 4) that regulated the development of REDD private projects. Instead, these funds were part of Mozambique's application to the World Bank's FCPF, created within the UNFCCC framework to prepare developing countries for REDD.

The second phase of REDD—already mixed with the new landscape approach—included US$44 million under a private investment fund called a Forest Investment Plan (FIP), US$46.3 million for the Conservation Areas for Biodiversity and Development Project, and US$40 million for the Agriculture and Natural Resources Landscape Management Project. Supposedly, this was to be followed by a third phase, in which emissions reductions would be verified and paid accordingly, up to the amount of US$50 million. It was not exactly clear how much of these funds had already been disbursed at that time—except for the first US$8.6 million—nor who and how the length of these phases would be determined.

In both my interlocutors' conversations and the World Bank material it became almost impossible to distinguish between what exactly was considered REDD and what was part of the landscape program. This blurriness around REDD and this new, more holistic perspective seemed to constitute a form of purposefully dissolving REDD into numerous activities that were supposed to reduce deforestation in the jurisdictions targeted by the program.[49] So, for instance, in November 2016, the Mozambican government approved its national strategy for REDD (2016–2030)—an official document demanded by the World Bank as part of the REDD process,[50] but all six of the strategic goals approved in the document are exactly the same ones found in two World Bank brochures describing the landscape approach in Zambézia or Mozambique in general.[51]

The strategic goals of the approved National REDD Strategy include the constitution of an institutional and legal platform to coordinate all the activities leading to reducing deforestation (a goal embodied in the constitution of the UT-REDD); conservation agriculture (as a means to reduce slash-and-burn practices); improved biomass energy (rendering charcoal production more efficient); strengthening of conservation areas; sustainable forest management; and the restoration of degraded forests and reforestation. In the case of the Zambézia province, these goals are supposed to be implemented under the landscape program that covers nine districts, with a total area of 53,000 square kilometers. It was during an official presentation of the government's plans for Zambézia that I was made aware of how REDD was gradually being swept under the rug—at least at the level of public discourses.

In April 2016, members of the UT-REDD and from the World Bank traveled to Quelimane, the capital of Zambézia province for a two-day event with the Provincial REDD Forum. This Forum was created as part of the legal and institutional framework demanded by the World Bank during the REDD "readiness" phase and integrates local government officers, farmers, loggers and forest operators, local NGOs and companies, staff from the Gilé National Park, and some *régulos* (traditional authorities). I was allowed to travel from Maputo with the whole delegation to attend the forum. We were joined at the airport by two Brazilian men, from a Brazilian NGO that had been commissioned by the World Bank to draft further documentation related to the landscape program. To my surprise, one of these men was Celso, the founder of this NGO, who I had interviewed over Skype while still in Acre, due to his NGO involvement in a REDD project in indigenous lands in the Brazilian states of Rondônia and Mato Grosso. It was the first time that we were meeting in person, albeit in Mozambique, not Brazil.

Celso turned out to perform the role of mediator and coordinator during many of the debates throughout the entire event, and frequently met with Arnaldo (the World Bank team leader) and Manuel (the leader of the UT-REDD) outside the forum proceedings. The event was held in a government building with a beautiful view of the bay and was intended to introduce the landscape program in a workshop, during which participants would provide their inputs. Right from the beginning, both Arnaldo and Manuel mentioned the milestones accomplished during past meetings[52] and the need to broaden the scope of the forum; hence the initial proposal to change the name of the forum to Rural Development Forum, "because this is not just about REDD anymore," said Manuel. As such, the event served the purpose of introducing all the dimensions of the landscape program. While these dimensions were at times connected with the need to reduce deforestation, the goal of the whole program was "rural development." REDD was not mentioned, neither as the means nor the end.

When I first met Arnaldo, in Maputo, before I could even complete the introduction to my research, he interrupted me by saying: "REDD is dead!" To which I thought: *dead even before it existed.* ... During the forum, one could almost agree with the assertion that REDD had indeed died, if it were not for the fact that the World Bank funds that Arnaldo was presenting only existed and were being disbursed because of REDD. In other words, while that event and some of the activities mentioned during it were being financed under the REDD umbrella, at the local level there was a clear effort—led by both UT-REDD and World Bank officials—to remove REDD from their discourses.

FEELING DÉJÀ VU: CONSERVATION AGRICULTURE

As in Acre, in Mozambique too, small farmers using fire to clear fields are considered the main culprits of deforestation—even if other evidence points to more important causes such as urbanization or the expansion of commercial agriculture. Although the assumption that slash-and-burn practices are necessarily associated with higher emissions levels is contested in the scientific community (see Ziegler et al. 2012), environmentalists continue to demonize the use of fire in agriculture, suggesting conservation agriculture as the solution. Accordingly, a French NGO that has been working in the Gilé National Park developed a project for the communities around the park with the goal of increasing crop yields, as well as reducing the pressure exerted by smallholder slash-and-burn practices on forests inside the Gilé Reserve.

During my conversation with Pierre, one of the founders of this NGO, he explained that besides preparing a small REDD project inside Gilé—at that point seeking certification from the VCS—the NGO had developed efforts to engage local communities by offering training in conservation agriculture. The main reason for these efforts, he said, was the involvement of these communities in illegal logging inside the reserve.[53] I asked him more about the illegal activities inside the park and while Pierre confirmed that it was a big problem, he also added that this type of illegal logging did not configure deforestation, but instead, degradation. Since the individuals involved in this activity seek to steal *pau ferro*—a highly valued type of timber[54]—leaving all the rest untouched, this activity configures forest degradation and not deforestation.[55] However, this degradation harms the integrity of the REDD project they are seeking to certify. Notably, this REDD project includes "sustainable forest management," not to extract timber but to facilitate tree restoration, promoting the growth of new trees, clarified Pierre. This is the same argument used by authorities in Acre to associate "sustainable forest management" with REDD. Ultimately, this argument also shows how the difference between sustainable forest management and illegal logging seems to merely rest upon a legal stance: while the former entails removing only the older (and thus more valuable) trees to promote forest restoration, the latter involves cutting only the most valuable trees (i.e., the older ones), but without a legal permit to do so.

Pierre confessed that it was hard to implement conservation agriculture, since Mozambicans are "risk averse" and extremely "resistant" to changing their practices. This same resistance to change was mentioned by Arnaldo as the reason why it will be so difficult to reduce deforestation in Mozambique:

> They do the itinerary system [slash and burn] because it's good, it's how they have always done it, and because there is no support to do otherwise. To change that is very difficult. Really very difficult! Because we are talking about changing the behavior of how many million people? In Zambézia alone, you have more than 500 thousand producers. We're talking about having each one of them stop doing agriculture that way and to do it differently. Now, honestly, if you want to reduce deforestation, you can forget everything, just do one thing: agriculture! You don't even have to mess with charcoal [production]. Mess with agriculture. If you don't change agriculture, forget about it. If we can't really enhance the productivity of small farmers, and with the birth rates here in Mozambique, forget about it, there's no way.[56]

Following the same logic expressed by Arnaldo, Pierre's NGO is trying to "mess with agriculture" by paying farmers a higher price for their cashew and sesame seeds in exchange for their adherence to conservation agriculture. According to Pierre, all families have cashew trees, but since market prices are so low, they do not bother to process the nuts. The World Bank wants to change this as part of the landscape program, he added, in which cashew and sesame seeds are elements of the "green supply chains" to be promoted, along with a greater orientation toward the market. Arnaldo had explained that the landscape program includes the "integration of small and medium emerging farmers in value chains" instead of subsistence agriculture. But what does that mean, I asked him.

> It's very simple. It's access to credit, access to mechanization, access to technical assistance . . . it's all of that that we're trying to get now, with these US$80 million from the Bank. So I think that that REDD thing has to stop tripping on carbon and so on, and enter the rural world, which is way more complex.

I was struck by the extent to which the language used by Pierre and Arnaldo when referring to cashew and sesame seed production was similar to the language of Acrean officials describing the "productive chains" of nontimber products, like the Brazilian nut or açaí. In both cases, these "value chains" were being addressed exclusively from the supply side, ignoring the demand, not to mention the competition from other supply sources. It seemed that for Pierre and Arnaldo, to increase Mozambique's cashew production alone would be enough to change the entire international commodity chain, turning the country into a major world cashew producer and exporter.

Arnaldo's claims about the plan's simplicity aside, access to credit turned out to be problematic in practice for most target beneficiaries. During the Zambézia Forum, Manuel explained how the credit program would work: each proponent (what they called an "emerging" farmer) of an agricultural project involving twenty small farmers would have their project funded by the World Bank's grant in 50 percent, as long as the farmer owned 10 percent in cash, and sought the remaining 40 percent in commercial bank loans. Even though Manuel was addressing a sympathetic audience, protests emerged immediately: did Manuel have any idea how high banks' interest rates were? Did he not understand that no farmer has "ten percent of anything in cash?" people asked angrily. Troubled, Manuel turned to Arnaldo who came to his rescue, saying that commercial bank loans were not mandatory—people could have enough on their own to conduct their projects—that 50 percent of the money was actually given away by the World Bank, and interest rates from commercial loans could actually decrease with the World Bank's guarantees on the projects. In fact, he added, interest rates would certainly go down in time, due to the involvement of the bank. And the 10 percent did not have to be cash; it could be equipment or other assets, he suggested. Although people seemed more appeased, the idea of resorting to commercial loans continued to echo throughout the forum as something almost impossible to achieve.

Before watching this plan being presented to an audience, I had the chance to chat about it with a member from the UT-REDD who I met casually during one of my visits to that office. He told me that the plan consisted of helping out an emerging farmer (meaning a farmer who already produces for the market) to expand his production, under the assumption that by elevating his capacities to supply markets, this emerging farmer will help other small farmers around him by buying their production, so he can respond to the increased demand of the markets. The plan presented at the forum consisted of granting these emerging farmers with increased resources so they could expand their production (see box 7). Once again, this vision was focused exclusively on the supply side of the markets—thus ignoring other conditions of markets' functioning—while also trusting in the power of the markets alone to bring development to Zambézia. This belief in the market's power to elevate farmers' livelihoods had already been expressed to me very bluntly by Manuel:

> We're not giving away seeds. We'll bring someone who sells seeds. And the farmer will do business with him. Or we'll teach the farmer that there is a store selling seeds. And he buys. So we're creating a link between the farmer and the seed seller. He produces, [therefore] his production needs to be sold. We'll show him that there is a market. We will not carry him like a baby anymore.[57]

Setting aside the assumption that Mozambican farmers need to be taught about seed stores or the existence of markets for their produce, Manuel seemed to ignore the fact that access to markets is a complex issue, since Mozambique continues to face serious problems of mobility to and from rural areas. However, more important than actually achieving the goal of increasing the role of emerging farmers in the development of Mozambique's rural areas was the fact that the World Bank was seeking to expand a market-based development model. In the process, World Bank officials are able to inform national policies and governing structures toward more market-friendly approaches, exercising a form of subjectivity that can have enduring effects in the country's policymaking.

> **Box 7**
>
> **From *cantineiros* to emerging farmers**
>
> The types of agrarian relationships that the World Bank is seeking to promote resemble the hierarchical structures created by the chartered companies in the areas under their control. Such is the case of a development plan supported by the World Bank in which the success of small farmers is predicated upon the financial support of those who are already more successful, following the assumption that the wealth of these successful farmers will necessarily trickle down to the poorest ones. In practice, this type of agrarian structure is very similar to the colonial network of *cantineiros* (small traders who collected the production of small producers on behalf of colonial authorities) and peasants, around which agrarian relationships were organized in Zambézia.
>
> There are also continuities in the rationale underwriting the methods and choice of crops by Mozambican authorities and the World Bank to implement "rural development"—the priority given to industrial plantations over family farming and to monoculture over risk-averse and diverse crops is clear (despite the long history of unsuccessful development projects), as is the election of the same crops once promoted by colonial authorities: cashew, sesame, cotton, and tobacco (Isaacman 1996; Temudo and Silva 2011; Penvenne 2015).

The most remarkable element of these discourses about commodity chains by Manuel, Arnaldo, and Pierre was the feeling of déjà vu that they all entailed. After a disastrous intervention by the World Bank in the early 1990s that collapsed the entire cashew sector in Mozambique, the bank (and the government) seemed ready for another round.[58] This time, incentives to cashew production were tied to wider efforts to reduce deforestation—both by implementing conservation agriculture and by using cashew trees (now imbued with a greater value) as deterrents to the use of fire. But the historical repetition was embodied as well in the assumptions that conservation agriculture—an element of "development" interventions in Mozambique since at least the 1990s—is the most important instrument to fight shifting agriculture and improve agriculture outputs for markets.

The realization that these repetitions arrive with no new nuances led me to express my frustration days later with a Mozambican writer, with whom I was exchanging perspectives on Mozambican politics over drinks. In a facetious way I said that Mozambican farmers must be really stupid since they have been taught conservation agriculture for more than twenty years, and still continue to not know how to do it. The writer shook his head, sharing my frustration, and added that these NGO practitioners simply do not understand that farmers cannot take any risks in agriculture since their survival is on the line. More to the point, what these historical repetitions show is that the REDD template can be invoked to repeat—all over again—many experiences from the past, all of which failed to achieve their stated goals. The only difference is that these same activities are now tied to the goal of reducing carbon emissions.

HIGH-TECH CHARCOAL

In the materials produced by the World Bank and the Mozambican government related to REDD and the landscape program, charcoal production was also identified as another main driver of deforestation.[59] In the approved National REDD Strategy, it is explained that this situation is due to Mozambicans' low incomes and lack of energy alternatives. Therefore, although there is electricity in urban areas like Maputo, its high cost and lack of infrastructure for other energies (such as gas) leads people to use charcoal to cook. Gas stoves require a high up-front investment that most families cannot afford, and gas supplies do not reach all urban areas (let alone the rural ones). The same document states that in 2011, more than 80 percent of the energy used in Mozambique originated through biomass, and it cites a study stating that all the charcoal used in Maputo comes from Gaza and Inhambane (the closer two provinces) because the forests in Maputo province have already been entirely depleted (República de Moçambique 2016, 24)—this citation is provided as evidence that the remaining forest areas of the country will have the same fate if nothing is done to prevent it.[60] The policies to be implemented to address this problem do not include providing low-income families with access to gas. Instead, the goal is to make the use of charcoal sustainable. The strategy states that, although charcoal production is a driver of deforestation, biomass is also a renewable source of energy—as long as it is produced "sustainably," from "sustainably managed forests" (24).

The landscape program in Zambézia includes the distribution of efficient cookstoves—which supposedly reduce charcoal consumption—training local charcoal producers to use kilns with high levels of efficiency and planting rapid-growing tree species, such as eucalyptus.⁶¹ The most striking point about this plan is how it blatantly reproduces "a series of misplaced assumptions about household or domestic energy, deforestation, and biomass fuel" held by international development agencies since the 1970s (Crewe and Harrison 1998, 11). According to these assumptions, the use of efficient stoves would necessarily decrease deforestation. The accumulated experience of efficient cookstove distribution has instead showed that adherence to the stoves is dictated more by interest in saving time than in saving fuel, and that most distribution programs resulted in increased gender inequality and greater benefits to stove producers rather than those who were supposed to gain from using the stoves (Crewe and Harrison 1998, 11–14).⁶² Despite a history of contested success, Manuel was excited about the cookstoves, and especially about his ideas for sustainable charcoal. He detailed his plan to me:

> I was planning a very interesting project like this: you would get three or four people in the districts that would go on to collect the leftovers from wood concessions. You know there is wood, and that they only take [the good one] leaving leftovers. They would collect all that and deliver it to those who process it, putting it in bags, and then they would sell back the charcoal to those guys in bicycles. Those guys in bicycles buy directly from the factory to sell in markets. So you would be doing two [actually three] interesting things: you're using the leftovers from the wood concessions, you're giving a job to those collecting the leftovers, and you're giving a job to those who are going to sell the wood. A fourth point is that you're adding value to your charcoal. Because it is a charcoal that is packed, produced sustainably, and sold without you getting your hands dirty, at a good price.⁶³

What this plan fails to account for is how poor Mozambicans might be convinced to be only a part of a production chain, when normally they control the entire process from wood collection to sale, through household labor. Moreover, this plan seems to worsen the already existing problem of small-scale producers' marginalization: large-scale charcoal operators curtail access to both wood resources and charcoal commercialization rights, depriving many rural households of a crucial source of income (Baumert et al. 2016).

Ultimately, and despite the long history of unsuccessful curtailment of deforestation through the use of improved cookstoves, their distribution was yet again reintroduced as a new instrument to reduce carbon emissions.

This solution consisted of a blunt repetition of history by the World Bank. In a 1990 publication about the Norwegian aid to Mozambique, the authors already noted the bank's misconceptions of both the diagnosis and the solution to the problem of deforestation:

> Deforestation has been identified as the main environmental issue, with logging, fuelwood collection and clearance for agriculture identified as the major causes (World Bank, 1988). . . . This World Bank review presents a gloomy picture of the state of Mozambican forests, yet it would be a great exaggeration to interpret this as a generalised catastrophe (16).
>
> There are several indicators that suggest deforestation is not as severe as elsewhere in the region. . . . In fact, charcoal plays only a small part in fuel supplies in the Mozambique fuel market (SADCC 1989).
>
> The World Bank is probably unduly optimistic in expecting to achieve much substantial conservation through improved stove design quite simply because, with a plentiful wood resource in the country as a whole, there is little pressure to conserve wood except in areas of scarcity. (Cherrett et al. 1990, 17–18)

YOU WILL GET MONEY FROM THIS ELEPHANT FOR A LONG TIME

By linking deforestation with loss of biodiversity, the approved REDD Strategy asserts the importance of preventing the degradation of existing conservation areas through illegal logging, charcoal production, and slash-and-burn agriculture. As such, the document refers to the challenge of sustainably managing those areas in the face of dwellers' activities considered as threats. While the strategy points to the need to increase the sources of income to maintain the conservation areas and provide alternative livelihoods to those dwellers, there is not much detail on how those sources of income will be distributed to the communities. Instead, emphasis is placed on the need to "promote and introduce, in the local communities, activities that are compatible with conservation, and to establish an education and training system oriented towards conservation" (República de Moçambique 2016, 27). These compatible activities are mostly tourism, as Luís explained to me:

> If you maintain the biodiversity, you can attract ecological tourism. So, instead of killing the elephant, killing the gazelles, killing this or that, no. You

keep this and you will use it in a different form. If you kill today, it's finished. Tomorrow you'll have no money. And you need to see, there will come one hundred tourists to see the same elephant, so you will get money from this elephant for a long time. If you kill, you will have it today. Finished. Tomorrow you will have nothing. That's the spirit.[64]

The vagueness of this proposal was apparent during the Zambézia Forum, when a member of the government introduced plans to develop a tourism project to promote economic growth and poverty reduction of the communities around the Gilé National Park. While she insisted on the need to involve "local communities" in the "sustainable management of natural resources," during the debate she never answered questions regarding compensation to families that no longer have access to the park's resources and on which their survival depended. She was not able to justify either why the park's ombudsman is always in Maputo and never at the park, where people need him. Moreover, she never explained how those communities would "get money from the elephants," as Luís had suggested during our conversation, since the sharing of benefits from the park was not a topic of discussion. Given the emphasis of government discourses on the importance of tourism to sustainably maintain conservation areas, it is fair to assess that the US$46.3 million from the World Bank would likely be spent in reinforcing current business-oriented models of conservation (Diallo 2014, 54) that have mostly reproduced colonial logics of spatial organization (Hughes 2006), increasing existing processes of land dispossession (Lunstrum 2008), and forcing people living inside conservation areas to relocate (Lunstrum 2010).[65]

This goal of biodiversity conservation through the promotion of ecological tourism is not new, nor was it introduced in novel ways. Instead—and like all the other elements of the REDD Strategy and the Zambézia Landscape Program—was being discursively reframed to match the purpose of reducing carbon emissions.

FUTURE SUCCESS OF PAST FAILURES: COMMUNITY FOREST MANAGEMENT

As in the discourses I found in Acre, authorities in Mozambique kept repeating that logging must be done in a sustainable way to avoid carbon emissions. Following that logic, the REDD Strategy approved by the government identifies a series of measures toward training wood dealers and their certification, improving their connections to markets and to financial institutions, and improving state control over wood concessions (República de Moçambique 2016, 28). Although during my conversations with people from the UT-REDD and from the Environment Ministry there was a strong emphasis on the need to promote community forest management, the subject was not mentioned during the Zambézia Forum, and there is only one item in the strategy specifically referring to this issue: "To enhance the economic integration of wood concessions and communities and ensure the formalization of private-public-community partnerships in the co-management of forest resources" (28). As such, this document confirmed the fears of some activists and government officials who considered that the Landscape Program did not have in consideration the interests of rural communities.

Community forest management is not a new thing in the history of rural Mozambique. It was implemented in the country during the early 1990s, right after the end of the civil war, and following international trends of allying environmental conservation with local (and indigenous) communities.[66] I talked about the failure of this experience with many of my interlocutors, trying to understand why community forest management had failed, why such a failure was being revived in the context of a supposedly new approach to landscape management, and why government and World Bank officials were championing a past failure as a future success. During a long conversation with agronomists from the Agricultural Research Institute of Mozambique, I was told that the Mozambican experiment was an attempt to copy similar initiatives in Zimbabwe and Namibia. But while in these countries there was a strong touristic component (due to the existing fauna) providing more and faster revenues to the communities, in Mozambique there was a systematic lack of proper governance. Ultimately, the success of the projects relied on the commitment of individual community leaders who were not properly supported by the state.

Sofia, a member of a Mozambican NGO that was deeply involved in these communitarian experiments had a more critical perspective:

> On one hand, I say that community leaders have the power to manage the resources from their areas based on customary law, but when we're talking about profitable activities, I immediately remove from these leaders their power to decide.[67]

Moreover, she added, "we have always kept communities under a situation of dependency," "dependency towards NGOs, towards the State, towards donors, and so on." In sum, Sofia concluded, there was a total disconnect between these experiments and the rest of the government's policies. A member of the government[68] assumed a more denunciatory position regarding authorities' role in the failure of community management. According to him, there were community forest management projects all over the country implemented by NGOs and the Agriculture Ministry and strongly supported by international partners. However, these projects implied the devolution of power to rural communities, so the government invested in their failure, sabotaging them, because it could never allow such a significant decentralization of power. The government's resistance to decentralization has its roots in the violent conflict that divided the country after decolonization; therefore, such decentralization of power to rural communities might have been perceived by the government as too risky, given the large influence of RENAMO in many rural areas (see box 2).

Members of the UT-REDD and the World Bank sustained a different perspective on these past experiences. Confronted with my question on why something that failed in the past would work now, members of the UT-REDD and the World Bank provided vague answers explaining failure with poor management infrastructure, the small scale of projects, and their lack of orientation toward the market to ensure long-term sustainability. The deficient connections to the market were especially pointed out by Manuel and Romeu as the main cause for the failure of these projects—an assertion that is not surprising given their past experience in the N'hambita project.[69] All these problems had, according to them, been resolved with the landscape program, which addressed all previous shortcomings: the implementation of the program had begun with the creation of a comprehensive management infrastructure—from the UT-REDD in Maputo to the provinces fora, promoting an extensive participation—it coordinated all existing small projects into a larger structure, and it targeted markets. In the end, the "communitarian" part of sustainable forest management was more of a rhetorical device than a real objective to be achieved within the whole program—such a device has become almost mandatory in all development projects and is a crucial element to gather political support in development interventions.

In the end, the future success of this communitarian initiative was merely based on its discursive construction: members of both the UT-REDD and the World Bank had constructed a compelling narrative about Zambézia's potential, identifying a series of activities that would muster that potential in order to solve all of its problems while promoting rural development—a narrative that was effective at enrolling a large number of people, from international donors and NGOs to national authorities and local beneficiaries (Mosse 2005).

PLANTING EUCALYPTUS SUSTAINABLY

Although the national strategy on REDD does not mention the FIP, its sixth goal entails the plantation of one million hectares of forest by 2030 (República de Moçambique 2016, 28)—a goal that cannot possibly be achieved through small scale agroforestry systems like the ones that supposedly are promoted along with conservation agriculture and community forest management. Such a reforestation target necessarily implies industrial plantations and the FIP promotes exactly that. Introduced as a strategy to strengthen the forest sector, reduce emissions, and promote rural development, the FIP arose to engage the private sector in these goals, scale them up, and "achieve transformational change" in Zambézia and Cabo Delgado (República de Moçambique and FCPF 2015, 7). However, the FIP's first draft states that the investment plan intends to capture additional private investment "to eventually cover the entire country" and "to provide a holistic approach to the reforms of the forestry sector and the implementation of the national REDD+ strategy across the country" (51). This draft repeats elements from the National REDD Strategy,[70] such as descriptions of Mozambique's forests, drivers of deforestation, and solutions to curtail it—but what the UT-REDD and the World Bank emphasize as the most important element of the FIP and the Landscape Program in general is its engagement with the private sector through a subcomponent "focused exclusively on a major forest plantation client" (54), that is, a Portuguese paper and pulp company.[71]

This investment is described as a mosaic plantation of eucalyptus interspersed by houses, agricultural fields, and conservation areas. To this effect, the government ceded the DUATs (the right to use the land) of 356,000 hectares to this company, of which two-thirds will be planted with eucalyptus. The document states that within the DUATs there are 25,000 households, and while this investment is hailed for creating "a significant amount of casual labor" for those households as wells as other benefits (like roads, bridges, communication infrastructure, schools, and medical facilities), it also admits that sustaining the mosaic approach in the Mozambican context can be a challenge.[72] This was clearly mentioned to me when I met Matos, the president of the Portuguese company, in his office, located in a beautiful house with a garden, in the most expensive part of the city. Referring to the arson events suffered by other companies, Matos told me that his company's investment can only work if the plantations do not force people out of their lands. That is the reason, he explained, why his company asked for such a large extension of land, with enough space to maintain the existing households while planting eucalyptus around them, with their agreement.[73] According to Matos, the biggest obstacle to the project is the pervasive poverty of the region, with more than 60 percent of families earning less than US$50 a year. For that reason, the company intends to provide jobs (mostly seasonal), and a social development plan mediated by an NGO. This plan includes support for agriculture, establishment of local infrastructures, improvement of school attendance, and a grievance mechanism.[74]

During our long conversation, I realized that not only did Matos not understand what REDD is, but also his company's integration in the Zambézia Landscape Program was very recent. After I explained how REDD is supposed to work as a mechanism to avoid deforestation, and that it has more recently shifted to include sustainable forest management and tree plantations, he concluded that his investment was therefore eligible to do REDD—at least in 10 percent of the whole concession, he specified. Moreover, since eucalyptus is a fast-growing species, it is very effective at sequestering carbon—he added—a statement that I also found in the FIP document justifying the investment in the plantations to reduce emissions. He recalled that it was the World Bank who invited him to be part of the Zambézia Program; however, he was not yet completely aware of how things would work.

Matos's reasons to invest in Zambézia were not, however, aligned with the government's concerns with high deforestation rates and poverty levels in that province. For Matos's company, Zambézia simply met its basic needs: good areas for forestry production, access to export routes (through the port of Quelimane), availability of electricity and water, and the possibility to release effluents. According to him, these effluents are not a health hazard—since they comply with European requirements regarding pollution—but have "a very strong visual impact."[75]

Matos was not present at the Zambézia Forum, but his company sent a representative: the person responsible for the social development project of the company, a man with a long and disenchanted experience in the development industry, who had recently converted to the private sector. In his interventions, he emphasized that although the company held the DUAT of those areas, it was still respecting the customary uses of the land, preferring to negotiate with families their adherence to eucalyptus plantation. During the debate period, some people accused the company of forcing families out, not directly, but because eucalyptus were being planted in their farming areas. They also questioned him about eucalyptuses' appetite for water and the risks involved in planting these trees near water sources. These questions infuriated the man, who called his interlocutors "ignorant," cautioned them to do their homework before asking such outrageous questions, and defied them to identify sources of water where eucalyptuses were being planted or families that had to move because of eucalyptuses. His outrage successfully impeded any questions from being answered, while the mediator of the debate called everybody's attention to the fact that the forum was about "sharing ideas and not about crucifying other participants."[76] By the end of the forum, the company's representative had been able to seduce some participants, scheduling a visit to the eucalyptus nursery and the company's headquarters, in an effort to involve Zambezian "emergings" to start their own investments in eucalyptus plantation.

The participation of this company in the Zambézia Landscape Program—although announced as a major new component of the program—seemed to be more the result of a coincidence between a private investor's agenda, the government's development plans for Zambézia, and the interests of the bank in disbursing funds, than of a planned environmental policy to reduce deforestation. Matos's company had already requested a loan from the IFC to expand its eucalyptus plantation and paper production. This investment had no environmental purposes, nor was it integrated in any government-led initiative to fight poverty. It was not a novelty either, since this type of plantation follows from a historical pattern since colonial times (see box 8).[77] The social component of this investment had been conceived by the company in an effort to prevent the tensions and violent conflicts that affected similar investments in other parts of the country, namely in Niassa (FIAN 2012). However, both government officials and members of the bank associated this large (and potentially contentious) investment to a larger environmental and developmental program, seeking to render it more appealing, less controversial, and, importantly, more able to capture additional investments. It is important to note that as in any other bank, the World Bank obtains its profits by conceding loans to developing countries, and that frequently, the goal of providing loans—even in the face of high fiduciary risk—is almost an obsession of the several programs managed by the bank, which has, as such, created a "culture of lending" (Berkman 2008). Even though Matos had no idea what REDD is, the premise that by planting eucalyptus at a massive scale the country can reduce its emissions was deployed by government and World Bank officials to legitimize and support this initiative within the Landscape Program. Even if taking in consideration the increased capacities of eucalyptus in carbon sequestration, all these planted trees will ultimately be cut down, as their only purpose is cellulose production.

Box 8

Planting eucalyptus to promote forest conservation

The increased demand for timber spurred by WWII generated some concerns among forest engineers working for the Portuguese colonial authorities regarding Mozambican forests' capacity to regenerate. For some of these engineers, the destruction of "primitive" Mozambican forests was due to the "unconscious action of [the native] man," using fire to clear fields and constantly moving into new areas (Cardoso 1946, 13). Although others deplored the uncontrolled expansion of cotton cultivation (Gomes e Sousa 1955) enforced by colonial authorities, or even the unsustainable actions of loggers (Esteves de Sousa 1950), the shifting agriculture practiced by Mozambican farmers was considered by far to be the main culprit of deforestation.

The solutions suggested by these engineers included the creation of forest reserves to promote natural regeneration (Cardoso 1946, 25), a more consistent zoning for forest concessions, and reforestation (Gomes e Sousa 1955). Most of this reforestation used exotic species (such as *Pinus* and eucalyptus) due to their fast growth and profitability (Comissão Técnica 1966), despite cautionary notes against the use of eucalyptus due to its impacts on soil erosion and water exhaustion (Gomes e Sousa 1955, 4). Notwithstanding the negative impacts of this exotic species, eucalyptus planting became a common practice in forest concessions held by South Africans throughout the 1990s (Hughes 1998; Chitará 2003) and, more recently, equally promoted by the World Bank as part of Mozambique's REDD program.

Finally, all the (vested) interests of this investment aside, the association of the eucalyptus plantation in the Landscape Program compounded an existing muddiness between the REDD national strategy and the Landscape Program, between public initiatives and private investments, and between government officials and World Bank staffers—all of which had important consequences for transnational governance (as I will explain in the final section).

REDD Is Not the Invention of the Wheel

During the Zambézia Forum it became very clear that REDD had been reduced to a mere footnote of a wider program, even if all activities planned or already implemented were justified under the rationale of reducing deforestation. In fact, it was amusing to watch Zacarias, a member of the UT-REDD, during a group activity expressing his suspicions toward REDD and carbon markets in general. Unlike Manuel and other members of the UT-REDD, Zacarias is an older man with a very large experience working in development. His comment was immediately followed by a woman who condescendingly said:

> We don't see REDD as a [financial] compensation, but as a bonus, because what we want is to change mentalities. REDD has created too many expectations, and nobody earns money with REDD. REDD is an opportunity to deploy legal mechanisms of supporting [that change in mentalities]. The reduction of deforestation can be paid for, but the market has not been established yet.[78]

Her words expressed an important functional element of REDD: while unsuccessful in constituting itself as a market, REDD has, however, produced important subjectification effects (Dempsey and Suarez 2016). It has helped change mentalities toward a more neoliberal approach to the environment, in which market mechanisms and the decentralization of governance play an important role.

As Zacarias continued to express his doubts, his Brazilian partner from the World Bank became upset and felt compelled to defend the mechanism, arguing that REDD was successfully working, especially in Brazil. In the end, the Mozambican remained unconvinced, and later told me, as we walked together to the hotel, that he could not support things he could not understand, and REDD was precisely one of those things. This exchange was particularly significant since Zacarias had worked in the N'hambita project, with Manuel and Romeu, and therefore should be familiar with the principles behind carbon offsets. However, it seemed that his prior experience at N'hambita had not been enough to convince him of the merits of carbon trading—especially given its intangibility.

As the forum continued, the pressure exercised by Manuel and Arnaldo to change its name from Forum REDD to Provincial Forum became more and more clear—an institutional sign that things moved past REDD. Both explained the necessity of this change by recalling the diversity of issues discussed during that meeting. Continuing to call it Forum REDD was a form of restricting what could be done to promote Zambézia's development. The audience agreed. Later on, I learned that the UT-REDD itself was in the process of changing its name to International Funds Management Unit, and its mandate to one of coordination in the application of these funds.

This nomenclature change did not mean, however, that carbon did not matter anymore. It still does, especially for the World Bank, which continues to allocate funds to its FCPF, while insisting on the potential of trading carbon offsets as a mechanism to tackle emissions. On a local level, replacing REDD with broader terms, especially those tied with "development," is especially productive as it enables broadening the scope of interventions, and simultaneously renders those interventions more legitimate. Conservation agriculture, sustainable forest management or eucalyptus plantations are certainly easier to grasp than carbon trading, as it is easier to understand how these activities—unlike carbon trading—can promote development. These activities are not different from those being implemented in Mozambique since its independence; the only difference is that they are now tied to a goal of reducing emissions, so they can be supported by international funds. But again, the goal of reducing emissions is not explained to Mozambican peasants, as *they would not be able to understand*. Reducing emissions is simply what justifies the mobilization of international funds, and what promotes enrollment (and therefore, more funds) among international audiences—but the activities continue to be the same.

Romeu made that very clear to me when, on our last day in Quelimane, we sat together over lunch. He said that REDD is in no way different from what he learned in the university and has done throughout his life as a forestry engineer. In forest management we estimate the number of trees we can cut without compromising the percentage of vegetable cover, he explained. With carbon, it is the same thing: we have to define a cut value that does not jeopardize the carbon stock level. Emissions caused by deforestation are estimated at this point in 0.84 parts per million (ppm), and we want to go back to the 2004 rate of 0.58 ppm (that is the baseline). If we can control deforestation up to the threshold of 0.58 ppm, we will already be reducing emissions. So, it is exactly like forest management, he concluded.[79]

This was not the first time I heard people say that REDD was not much different from what they were already doing. I had heard similar claims in Acre, and in my first meeting with Luís, he stated that they were not doing anything particularly new in Zambézia:

> REDD does not have to come here to invent the wheel. We already have organizations in Zambézia suitable to work with the communities in conservation agriculture. But sometimes these efforts do not have the support they need to grow and generate the desirable impact. So we have created this Forum, where we all work, each in each area, but with the same goal of reducing deforestation. We don't have to invent [anything new].[80]

As in Acre, REDD in Mozambique was just continuing to do what was already being done, which, from the strict point of view of carbon's accounting, violates the additionality rule (i.e., no emissions are reduced). This contradiction in the rules of carbon accounting was justified numerous times by both Acrean and Mozambican officials and NGO practitioners with the argument that such activities were being scaled up more in quantity and in extension was being added to the scenario.

The fact that the Zambézia Landscape Program was just scaling up what was already being done—except for the plans of planting eucalyptus at a massive scale, carried out by the Portuguese company—was not a problem for donors either, even in the face of well-known past failures. When I met a representative from the embassy of Sweden and asked why his country was supporting the Zambézia program when all the funds to support the Mozambican forest sector had been suspended in the recent past,[81] his answer was surprising, but telling. The main virtue of the program was, in his view, its ability to capture private funding, thus being able to continue in the future. I reformulated my question, inquiring about the importance of funding a program that can be successful in its stated goals of reducing emissions and able to sustain itself without any further external funding. Once again, he pointed to the potential of the program to raise more private funds and capture more donors over time—independently of its ability to accomplish the purported objectives.[82]

Part of the potential offered by the Landscape Program, I understood later, was not just its move past REDD, but its capacity to integrate a large variety of activities, involving multiple actors, all of which had different goals and agendas. The extension of the program in scale and in scope enabled the enrollment of many different participants (at local, national, and international levels), with many different interests. Notwithstanding, the presentation of the Landscape Program still had to be generally framed around the goal of reducing emissions through forests—all the activities suggested would prevent deforestation, and the eucalyptus plantation would add more trees to the carbon budget equation. This message was particularly relevant to international audiences, especially when seeking additional funding to support the program. For national and local audiences, however, the program was presented as an initiative to promote rural development and to involve the private sector in that effort. Ultimately, the goal of promoting rural development was also appreciated by international audiences as a co-benefit of carbon emissions.

Notwithstanding, in order to present the Landscape Program as a legitimate approach to tackle both deforestation and rural development, government and World Bank officials had to constantly sustain and reproduce an underlying narrative about Mozambique's deforestation and its connection to poverty. According to this narrative, the main drivers of deforestation are slash-and-burn agriculture and charcoal production—both of which can be addressed with development interventions such as conservation agriculture and so-called sustainable charcoal. Once this narrative is accepted, almost any kind of intervention in the agricultural sector can be classified as reducing deforestation, including the planting of eucalyptus. That is why the Swedish representative explained to me that we cannot have reduced deforestation without rural development. In the short run, he said, you will have an increase in deforestation and in the areas used for agriculture. But in the longer run, there will be a reduction in deforestation because those areas will be the same—no more farmers' mobility, he concluded.

Conclusion

As in Acre, the REDD process in Mozambique shifted considerably from initial efforts to prepare for future transactions in carbon offsets into an extended program involving an entire jurisdiction and integrating different activities—all of which deemed capable of reducing deforestation, and therefore, of reducing emissions. The main difference between Acre's jurisdictional program and Mozambique's landscape approach was that while authorities from Acre sought to deploy REDD as a set of different policies incentivizing sustainable methods of production, in Mozambique, the acronym REDD and its association with carbon emissions were erased altogether. Instead, officials from both the government and the World Bank asserted the goals of promoting "rural development" while recognizing that the purpose of reducing emissions was something they kept to themselves. Events in Mozambique did seem to substantiate the statement expressed by Arnaldo that "REDD is dead"—in fact, REDD died even before it started.

Independently of the potential environmental benefits of REDD, Mozambican and Acrean authorities supported the mechanism for their own reasons—mostly for the opportunity to access important funds to promote development. In both territories, all the different elements of the Jurisdictional or Landscape Program constitute the repetition of activities already tried in the past and failed. Failure and success are all discursive constructions, as Mosse (2005) has pointed out, but even if these past failures are set aside, scaling up activities that were already taking place constitutes a flawed logic. For international donors, however, what mattered was that these activities were accompanied by policies and governing structures that opened spaces for broader transnational interventions. Moreover, all these policies and governing structures—whether effective or not in their stated purposes—produced subjectification effects, promoting different perspectives about the environment (Dempsey and Suarez 2016). All the while, these international donors can claim to their national constituencies and other international audiences that they are doing something to tackle climate change.

The inclusion of so many different elements and activities in these jurisdictional programs—part of what I call capillarity—produces yet another important effect: confusion. Confusion between which activities are new and which were already taking place; confusion regarding which actors are responsible for each activity; confusion about the relationship between these different activities; and ultimately, confusion between the stated and unstated goals of each activity. In the case of Mozambique, this confusion was further compounded by the purposeful entanglement of the approved National REDD Strategy and the Zambézia Landscape Program—both documents identified the same elements and goals, precluding their distinction both as intent and as practice. Finally, this messy scenario was made even messier by yet another purposeful muddling of the mandate and activities of government officials and World Bank staffers. Throughout my fieldwork I faced the difficulty of understanding who exactly was directing a process, who had made certain decisions, who had commissioned consultancies, who would pay the consultants, who was sanctioning final drafts, or who had chosen certain NGOs instead of others. I was not alone: some of my interlocutors complained that, very often, they did not know for whom they were working, or simply, to whom they should report.

In parallel to the vagueness of the language employed in the agreement for the REDD Framework (see chapter 3) and the equally vague proposals in both Acre's and Mozambique's REDD programs, confusion is also generated at the implementation side. This confusion, entanglement, and muddiness is an important technique of transnational governance, as it facilitates its interventions by simultaneously requiring its expertise, while rendering its participation not noticeable. In other words, places such as Acre and Mozambique lack the technical and scientific skills deemed necessary to implement REDD, which requires the intervention of multiple international actors; these interventions are then attributed to local authorities, rendering the participation of international actors invisible. The problem of climate change is particularly apt as a subject requiring specific technical and scientific skills. It is a complex issue, involving different fields of knowledge and competing political claims regarding scientific knowledge. However, the ways in which transnational policymakers deemed carbon trading as an appropriate solution to tackle this problem made the issue even more complex and, therefore, in need of experts with the necessary skills to implement the solutions.

Frequently, these international actors become a constitutive part of the governing structures they helped create—as in the case of the WWF-supported officer in Acre's IMC. Every time I walked by the offices of the UT-REDD in Maputo, I could not tell exactly who worked in the unit and who worked at the bank across the street; during the Zambézia Forum, elements of the unit and of the bank performed the same coordinating roles without any distinctions among them; officials from the Ministry of the Environment worked in different offices and talked about the UT-REDD as an "elite cabinet" developing activities that seemed to be completely disconnected from those of the ministry. The invisibility of the World Bank officials in the coordination (and even leadership) of the REDD process makes it particularly effective while national authorities enable the intervention of these actors (sometimes in the pursuit of their own goals), providing their actions with the legitimacy that is usually granted to elected officials.

Despite (or perhaps because of) the confusion and vagueness created by and through the multiple interventions of various international actors, and despite the fact that such confusion rendered REDD even more elusive (to the point of being declared dead), there is a very concrete result produced by the implementation of REDD in its variegated (and at times contradictory) formats—the construction, authorization, and reproduction of a larger narrative, according to which the reduction of global emissions can (and should) take place in the Global South. More importantly than pressuring countries from the Global South into accepting market mechanisms involving their forests is providing legitimacy to the idea that these countries can be active participants in mitigation efforts. This narrative not only erases the historical responsibility of industrialized countries, thus nullifying their political accountability, but also shifts that responsibility to the Global South. Simultaneously, as the REDD template has shifted to include different activities considered to have the potential to also reduce emissions, the participation of the Global South in mitigation efforts is no longer confined to their forests either. It has expanded to include many activities, from agriculture to cattle ranching, from fish farming to ecotourism, from sustainable forest management to planting eucalyptus—all activities involving land.

Ultimately, the REDD template has enabled the expansion of transnational governance in more capillary ways than previous global interventions. The technical and scientific expertise required for its implementation; the demands for document drafting following predetermined templates (frequently requiring mastering the English language); the constant rendering technical of very political issues; the policymaking processes and corresponding governing structures in conformity to international rules; the integration of various activities and their subsequent wrapping in an emissions reductions' discourse; and finally, the purposeful muddling of actions, actors' interventions, and their different purposes reflect such capillarity and are all (successful) techniques of these new modalities of transnational governance.

The efficacy of the REDD template can be partly explained by its capacity to be taken upon by different constituencies with different purposes—as already mentioned, the reasons for Acre's authorities and the Mozambican government to adhere to REDD were different from those that led international donors to continue to support it despite all its contestation. This persistent enrollment in REDD—for different reasons and by multiple actors—is thus a crucial enabler of transnational governance, even if REDD is gradually replaced by some other acronym or terminology. Such transnational governance is already enabled and authorized by the central idea that spurred REDD in the first place: that global carbon emissions can be mitigated (cheaply and politically effectively) in the Global South.

Finally, the other important element providing legitimacy to these new modalities of transnational governance is their anchorage in the idea of a common global space—the planet's atmosphere—and the need to save it for everybody's sake. Since climate change is a global problem affecting the entire planet, the notion that an intervention is required to solve it can hardly be challenged. Unlike efforts to spread democracy or enforce human rights at a global scale, the problem of climate change has been defined and recognized as a scientific matter—in this sense, the international mechanisms and agents who defined the problem of climate change as such can be considered a global "anti-politics machine" (Ferguson 1994). But even if the need for an intervention to solve the problem of climate change is not questioned, the object and nature of such intervention, as well as who will be involved, can be challenged. The efficacy of the REDD template in the terms I have described above neutralize (or at least weaken) such challenges: the narrative that has been gradually asserted through REDD's implementation define forests and the land sector of the Global South as the prime object of such intervention, and international experts as the main agents—even if their interventions can only be authorized by local and national authorities. Again, it is this entanglement of local, national, and international agents that renders transnational governance accepted and legitimate.

Despite the enormous efficacy of these new modalities of transnational governance—in the making since at least 1992—and despite the unchallengeable notion that the planet needs to be saved, emissions continue to increase at an astoundingly fast pace, and human beings in the Global South face the highest vulnerability to climate change impacts and unprecedented levels of pollution threatening their survival.

Conclusion

Throughout these pages I make two different major arguments. First, that the ways in which the problem of climate change has been defined not only oversimplified a very complex and multidimensional phenomenon, but, more importantly, that such simplification—which I called synecdoche—will necessarily preclude us from adopting the right solutions to tackle that problem. To the extent that climate change continues to be solely defined in terms of a failure to take into account the cost of emissions, the solutions will continue to focus on the implementation of market-based mechanisms that can "internalize" carbon in production processes. This is a political argument that calls for a more holistic approach toward the environment and that considers climate change—among other environmental problems—the product of a growing neoliberalization and financialization of the policy world (Harvey 2005). Tackling the problem of climate change will therefore imply rupturing with these logics and with the subjectification processes inherent to neoliberalism (Dempsey and Suarez 2016; Fletcher and Büscher 2017) and imagine alternative solutions outside of this market-based framework. It is also an argument against the utilitarian (and perhaps neocolonial) premise that puts the burden of tackling climate change on the Global South, and, in doing so, reproduces a long history of persistent inequalities while also jeopardizing the future of those who will suffer more dramatically the consequences of climate change.

Second, I argue that the ways in which the problem of climate change has been formulated enabled the emergence of new forms of transnational governance. Unlike more conventional forms of transnational intervention, these new modalities anchor their justification and legitimacy by deploying the idea of the global (as in the need to save the planet), have a broader and more capillary scope of intervention, and are more dependent on the mobilization of the authority granted by science. Transnational governance is certainly not a new phenomenon—its origins can be traced back to the creation of several international institutions and agreements after World War II—however, since the problem of climate change has been included as part of the international policy agenda, new modalities of governance have emerged. I thus contend that the transnational governance of the climate—which the several efforts to implement REDD in Acre or Zambézia constitute only one instantiation—presents important distinctions vis-à-vis conventional forms of intervention such as international development or humanitarian assistance.

The first distinction of these new modalities of transnational governance lies precisely in the global matrix of climate change. Since the climate is global, and climate change affects the entire planet, the transnational governance of the climate is especially legitimized as an intervention to save all of us; that is, interventions in the Global South spurred by the need to mitigate climate change are no longer about "developing" or "aiding" territories in need, but rather, about saving humanity. The premise that the planet (or humanity) needs to be saved not only opens the scope, but also the scale of possible interventions, while imbuing them with an unchallenged authority.

The definition of the problem of climate change as it stands in the international agenda is the result of a recursive co-constitution between climate science and the policy world. In other words, the entanglements between climate science and the policy world formed a large epistemic community—linked by shared forms of knowledge production and a shared set of values and norms—that reproduces itself through the practices and the authority provided by the scientific knowledge that it produces in a recursive process. This co-constitution between science and policy yields the second distinctive trait of this type of transnational governance. Even though all forms of governance are substantiated by some level of scientific knowledge, the transnational governance of the climate is more dependent on leveraging the authority of science than any other forms of intervention. This dependence is due to, on one hand, the complexity of climate change itself, which requires the deployment of scientific discourses to explain and understand the problem in its multiple dimensions; and on the other, to the fact that climate science continues to be challenged by several parties who intend to discredit the problem and maintain their status quo. Even if the general public can equate climate change with global warming and with the impossibility of life on Earth, scientific discourses are still needed to justify the solutions being implemented.

The modalities of transnational governance enabled by the problem of climate change have also attained a level of capillarity that was never achieved by other forms of intervention. Such capillarity is manifested at the policy level, with the creation of legal apparatus, regulations, laws, and policies, and also the infiltration of foreign experts in the existing governmental structures; at the level of practices on the ground with multiple activities that involve interventions in areas other than the forest sector; and at the level of behaviors due to the subjectification effects of this type of governance. Moreover, and also tied to this level of capillarity, is the fact that transnational governance requires the active participation of national and local authorities instead of simply operating parallel to these authorities. As a result, not only the interpenetration and interfunctionality between multiple authorities becomes more profound, but also transnational governance is rendered almost invisible, or at least, its methods and effects become harder to locate within the activity of existing authorities. This level of intrusion and interfunctionality is similar to what occurs in interventions to spread democracy, the rule of law, or human rights; however, these types of interventions do not achieve the same scope and level of capillarity that the transnational governance of the climate has enabled.

As I stated in the introduction, transnational governance is not a mere upscaling of a national government. In fact, transnational governance is currently more directly tied to the furthering of the strategic interests of certain economic groups and organizations than of national states, which makes it harder to locate and identify. Moreover, it is a form of global governance that moves across various jurisdictional levels—the international, the national, and the local—and that in fact depends upon the mobility of people, objects, and ideas across those levels. By tracing REDD's social life as it circulates from an international arena—where it has been imagined and conceived—to several sites of implementation, and back to its international point of origin, and by identifying the multiple forms in which REDD has been continuously reinterpreted, re-signified, and implemented, I have showed how transnational governance operates (using REDD as its specific instantiation).

Language is perhaps the most important instrument of transnational governance. Part of the transnational governance of climate change occurs during (and through) the annual climate COPs, under the UNFCCC. By examining the language ideology of both the performativity of these meetings and their written outcomes, I have shown how these negotiations enable the constitution, stabilization, and legitimation of a specific language, through which transnational governance exercises its authority. This language is, first of all, vague. It strikes a delicate balance between a consensual agreement about what needs to be done and the absence of binding actions toward those needs, while allowing for the coexistence of different imaginaries on what the decisions might mean. This language seeks as well to continuously render technical what are contentious and very political issues, while also stabilizing a certain version of the past—no longer open for debate—thus foreclosing possible futures. Finally, the language is punctuated by the tropes of *transparency*, *accuracy*, and *consistency*, which are part of the rhetorical normalization and political neutralization of the negotiations and the agreements. In the end, instead of considering these negotiations as failing, their absence of tangible results, vagueness, and political neutrality are instead a necessary condition of transnational governance by both asserting the UNFCCC's legitimacy to govern the climate and keeping all parties permanently engaged in the process of talking.

Another crucial mechanism of transnational governance is its capacity to generate an environmental apparatus within the existing legal structures of the jurisdiction under intervention. If, at the local and national levels, authorities may take upon REDD for reasons other than environmental concerns, their active participation in implementing it is still crucial for transnational governance. It is their participation that provides legitimacy to REDD by imbuing it with a local justification, a local need, and a local appearance. Myths of native origin and narratives about local efforts to make REDD possible are part of this resignification. Following this process, local and national authorities must put in place the required legal structures that constitute clear evidence of a transnational intervention, since these structures, although local, still conform to a given (international and standardized) template. This template is thus responsible for the production of a local environmental apparatus that not only shapes the legal and regulatory infrastructures of a territorial jurisdiction (as in the SISA law in Acre or the REDD law in Mozambique), but also produces subjectification effects, shaping people's perspectives on forests and the value of their conservation.

There is a more bureaucratic aspect of transnational governance, equally important for its effectiveness, that operates through document-writing. As I have described in chapters 4 and 5, local and national authorities are constantly producing documents related to REDD and targeting audiences at different levels to comply with international requirements, to provide evidence of compliance, to access funds, or simply to engage other parties. This document-writing process has important effects, as it provides legitimacy to local activities (at local and international levels), creates the justification needed by international audiences to continue to support REDD (namely by stating that it is successful at reducing deforestation), and it also produces subjectification effects, creating new ways of understanding forests and carbon. Mozambique was, for that matter, a case in point: while there were no REDD activities on the ground, the number of documents constantly being produced by multiple consultants and NGOs was confounding—to the extent that, at some point, it was difficult to understand the purpose (stated and unstated) of all these documents.

Both processes of creating an environmental apparatus and document-writing occur within a purposeful institutional messiness that also constitutes an instrument of transnational governance. The participation of foreign experts within governmental offices enables a level of interpenetration between governmental authorities and external entities that creates a purposeful blurriness between the competencies of one and the other. Ultimately, it is hard (if not impossible) to determine who is actually making decisions, what is the scope and goal of such decisions, or even how those decisions will be enacted. In the case of Acre, this institutional messiness was compounded by the oppositional interaction between state and federal authorities, in which, by trying to implement a carbon market in the state, the government of Acre openly challenged the position of Brazilian federal authorities within the UNFCCC negotiations. In the case of Mozambique, there was a purposeful confusion between what was REDD and what was part of a broader development plan for the province of Zambézia; and, importantly, between the competencies of the national government and World Bank officials. Such messiness reflects how local, national, and international actors negotiate their agency in the pursuit of their own goals; but at the same time, it also facilitates the intervention of transnational governance by rendering it invisible.

Finally, this form of transnational governance constantly presents itself as merely technical, thus constantly eliding the political stakes of REDD's implementation (Ferguson 1994; Li 2007). All those involved in REDD—be it in its formulation as a mechanism to reduce deforestation or in its multiple forms of implementation—deploy a science-based, technical, and standardized language. This type of language and concomitant procedures have the effect of deflecting politically contentious issues and erasing local specificities. Even the definition of forest, upon which REDD relies, already constitutes a form of standardization that imposes a very specific way of understanding forests and their value. The strict deployment of this carbon accounting language enables multiple (and surprising) reinterpretations of what REDD can be. Therefore, REDD becomes the template under which activities that were already taking place are further legitimized, and new ones are justified. The multiple activities developed under the REDD umbrella constitute an expansion of the scope of possible transnational interventions, which are no longer solely confined to forests, but are instead involved in the whole rural world (including, in the case of Acre, indigenous affairs). Some of these activities, such as conservation agriculture, cashew, or açaí production have a long history of past (and failed) interventions, but under the discourse of reducing carbon emissions, acquired renewed legitimacy.

Both in Acre and in Mozambique, REDD is being implemented by continuing, or scaling up, what NGOs or authorities were already doing, in the pursuit of their own fantasies of development. What changed with the emergence of REDD was that all those different activities started to be associated with carbon accounting and justified under the goal of reducing emissions. One could argue that in very much the same way market advocates claim that carbon markets internalize the externality "carbon," what people in Acre and Mozambique started doing was a discursive internalization of carbon. In other words, authorities, NGO practitioners, World Bank officers, activists, and environmentalists started to internalize carbon (either in their policy narratives, or by actually estimating it) to all the things that were being done.

The significant transformations that REDD has suffered since its introduction in 2005 have rendered this mechanism similar to any other form of aid development (Fletcher et al. 2016)—as it has been implemented in places like Acre and Mozambique, REDD could be almost anything. Such transformations could be interpreted as evidence that, like many other development trends, REDD too will fail (DeShazo, Pandey, and Smith 2016). Many have indeed declared its death, even those involved in its implementation. However, and despite these bleak prognoses, REDD continues to garner support from international donors and development agencies, even if the acronym is no longer used or is buried under new jargon (as in "landscape approach"). The reason for this continuous support arises from the potential inherent to the simple idea embodied by REDD: that we can reduce global emissions through forests in the Global South. Even if REDD disappears as such, this idea will continue to mobilize international support. REDD presents, indeed, a great capacity to be adaptable to almost any activity—and, as such, to be the purveyor of many different fantasies—enabling different actors to accommodate existing initiatives and agendas, and international agents to deploy REDD in the pursuit of their own goals, and in the expansion of transnational governance.

In other words, the rationale underlying REDD—although environmentally flawed—makes possible the articulation of the different interests held by international authorities, national governments from industrialized countries, and the Global South. Whereas poor countries in the Global South take this idea as an opportunity to access international funds and to expand their space for maneuver in the international arena, governments from industrialized countries trust that while using forests to mitigate emissions, more radical transformations will not be required from them. Such radical transformations not only would be expensive but would also have the potential to anger political constituents by demanding difficult changes in their lifestyles. Tellingly, the greatest achievement of COP26 in Glasgow (November 2021) was an agreement by more than one hundred countries to end deforestation by 2030 (Greenfield et al. 2021). Ultimately, this narrative about forests in the Global South as our salvation from climate change enables the perpetuation of a tremendously simplified perspective of the problem of climate change. It also enables the maintenance of an equally simplified narrative about global deforestation and its drivers.

As long as the main driver of deforestation continues to be attributed to small farmers who use fire to clear fields, the various international supply chains that exercise their pressure over forested areas will remain untouched. As will also the historical economic infrastructures built upon the inequalities between countries from the South and from the North, and whose roots are located in colonialism—the same inequalities that have shaped climate negotiations. Fighting deforestation by addressing international demands for beef, soy, or timber would be very complex, if not impossible, with the tools that policymakers are currently willing to mobilize, and it would certainly involve interventions in industrialized countries—something that would radically change conventional schemes of developmental intervention with unimaginable political effects. As it is, the narrative that it is possible to fight deforestation *in loco*, in isolation, and focusing on small farmers only, opens possibilities that a complex perspective on the drivers of deforestation would not. More specifically, it opens the possibility of eschewing industrialized countries' responsibility and accountability over climate change, while facilitating further interventions in the Global South.

POSTSCRIPTUM

Much has happened since I last visited Acre and Maputo. Two years after I last spoke to Roberto, I finally understood his despondency from the last time we met. He knew the money made available by international donors would not be enough to implement all the necessary policies to reduce deforestation, and he was also painfully aware that the environmental discourse informing some of these policies has backfired vis-à-vis those who have not felt much improvement in their lives and those who see sustainable forest management as a hypocritical endeavor in a state like Acre. His fears were substantiated. In October 2018, Acre's government lost the elections to a right-wing coalition that had run on the promise of developing the state by expanding cattle ranching and industrial agriculture. Prior to these elections, though, things seemed to be already falling apart. The condom factory spiraled toward bankruptcy, leaving dozens of families without a buyer for their rubber; and the fish factory has been struggling with damages caused by floods and high debts that might dictate its future closure (Venicios 2019).

In very similar ways, Mozambican authorities also knew that the funds disbursed by the World Bank were insufficient, although their hope was that such funds could capture additional investments. During our conversation, Manuel explained that the World Bank's money was only available to put together the infrastructure of the landscape program. "There is no money to implement anything," he said.[1] But in order to be able to receive private investments, authorities needed to follow the World Bank's lead and become ready for the markets. Amid a situation in which it was clear that the bank was leading the whole process, I asked several times about Mozambique's sovereignty. In my last chat with Romeu, he made it very clear: all of them at the UT-REDD (now the Unit for the Management of International Funds) were well aware that the bank's goal is to disburse as much funds as possible and that Arnaldo gets himself a paycheck and better career prospects every time funds are disbursed into the country. "Nobody is fooling anybody," he stated. But people like him, Manuel, and others also need their jobs, even if, as in Romeu's case, they do not believe that conservation agriculture will work or bring development. "Ultimately, Mozambique needs to be part of international markets," he added. "Otherwise, the country will never develop," Romeu concluded, shrugging his shoulders.

Recent events in Brazil and Mozambique continue to evince the flaws and contradictions of these policy approaches to climate change. The new federal government that emerged after the 2018 election in Brazil openly challenged any kind of effort to prevent Amazonia's deforestation. Jair Bolsonaro, the new president—a direct descendent and admittedly an admirer of the military dictatorship—started his mandate stating that Amazonia was open for business (Pearshouse 2020), sponsoring the illegal occupation of indigenous lands (Phillips 2019) and the acceleration of deforestation, throwing Amazon's ecosystem into what many believe to be a point of no return (Eisenhammer 2021). And while many leaders of the industrialized world did not spare words to criticize Bolsonaro's policies, the European Union settled a new trade deal with Brazil to increase the import of Brazilian beef and soy—thus, actually supporting Bolsonaro's intent of converting Amazonia into a giant reservoir of cattle, agricultural products, minerals, and energy (Nicolás 2021).

This same contradiction has equally marked recent events in Mozambique. While lamenting the tragic effects of the worst tropical cyclone ever registered in Africa, the Idai (considered a consequence of climate change) investments in the development of fossil fuels have not been curtailed. In fact, the French oil company Total has been developing a giant gas field in the Mozambican province of Cabo Delgado, which has also been the center stage of terrorist attacks in the region. Considered one of the poorest provinces of the country, Cabo Delgado is, however, very rich in mineral resources, attracting several international corporations. These companies are often responsible for the displacement and further impoverishment of local populations, fueling their resentment and turning younger generations into easy targets of radical Islam recruiters. Indeed, the terrorist attacks that killed nearly three thousand people since 2017 have less to do with the expansion of Al-Shabab in Eastern Africa than with the increasing poverty and inequality of Cabo Delgado (ICG 2021; Feijó 2021). Yet, French and American authorities are more willing to support a military solution that can protect Total and other foreign companies than in providing remedy actions to local poverty, while also refusing to address the fact that these investments only increase GHG emissions. In sum, not only are these investments responsible for the increasing violence in the region but they are also aggravating climate change.

This apparent dissonance between political discourses condemning environmental destruction and practices that actually promote such destruction (as well as social disarray) is a powerful demonstration that climate change cannot be tackled without a radical restructuring of the industrialized world's model of development. But until such restructuring is possible, the industrialized world will continue to plant trees while insisting on the responsibility of each person to reduce their carbon footprint.

Notes

Introduction

1. For more information on this event and its subjectification effects, see Machaqueiro 2019.
2. My emphasis. The transcription of this talk delivered by Naomi Klein, at Johns Hopkins University, was kindly given to me by my colleague Chloe Ahmann.
3. The first president of independent Ghana, Kwame Nkrumah, coined the term neocolonialism to refer to how former colonial powers and emerging superpowers like the United States continued to play a decisive role in independent countries. The idea of REDD being an instrument of neocolonialism is particularly apt in the context of discussions about political agency and national sovereignties within international policymaking frameworks.
4. For more critiques of REDD, see Leach and Scoones 2015, and Paladino and Fiske 2017.
5. Later, other researchers found that the Reinhart and Rogoff study had a gross mistake in the data input that dramatically changed their conclusions (Herndon, Ash, and Pollin 2013; see also http://archive.economonitor.com/lrwray/2013/04/20/why-reinhart-and-rogoff-results-are-crap/). Furthermore, the IMF recognized that the austerity measures should have been balanced with more Keynesian policies.
6. See https://www.theguardian.com/world/2018/apr/17/poland-violated-eu-laws-by-logging-in-biaowieza-forest-says-ecj.
7. See https://www.wri.org/blog/2014/07/tar-sands-threaten-world-s-largest-boreal-forest.
8. See https://www.nytimes.com/interactive/2018/08/01/magazine/climate-change-losing-earth.html.

9. The word *lusotropicalismo* links *luso* (of Portuguese origin) with tropicalism. According to Brazilian sociologist Gilberto Freyre (1933), the Portuguese were benevolent colonialists and "miscigenated" with native Indians and African slaves, thus originating a nation without racism. Unlike other colonial forces, Freyre argued, the Portuguese were able to create a civilization in the tropics, linking Portugal to Brazil and other African colonies. This imagined brotherhood is still part of the political rhetoric sustaining the Community of Portuguese Language Countries.
10. Since 2004, the Brazilian mining company Vale has explored one of the world's largest coal reserves in the province of Tete. This open mine has been widely criticized for its environmental impacts and for causing the forceful resettlement of 1,300 families into other areas. See Human Rights Watch (2013).
11. Clandestine settlements founded by escaped slaves that provided refuge to other escaped slaves.
12. Fieldnotes, July 15, 2014. For other accounts of how Brazilians feel they know Africa, see Cesarino 2017.
13. According to André Cicalo, this *blackness* is different from the *Africanness* deployed by Brazilian diplomats during the 1960s and 1970s. This *blackness* is deeply influenced by contemporary Brazilian black politics that have consistently challenged the myth of racial democracy and affirmed Brazil as the biggest black nation in the world after Nigeria (Cicalo 2014, 25).
14. Prosavana is a triangular agricultural project between Mozambique, Brazil, and Japan for the production of soy for exportation. Intended to shift Mozambique from a subsistence agriculture model into a commercial and extensive one, it replicates the "successful" model implemented in Brazil's *cerrado*. Including an area of eleven million hectares (the Nacala corridor) throughout nineteen districts in the provinces of Zambézia, Nampula, and Niassa, the dimension of the project, and the (false) claim that it would be developed in unused land, mobilized a series of protests and accusations of land-grabbing, forcing the government to reduce the project's scope. See Shankland, Gonçalves, and Favareto 2016.
15. In Portuguese, Movimento dos Trabalhadores Rurais Sem Terra (landless workers movement). The MST is a movement of people without land that advocates for land reform and distribution, fighting against several forms of inequity. The MST is considered the largest movement in Latin America, with over a million and half members.
16. Interview, July 3, 2014.
17. Peter van der Veer argues for maintaining the comparative approach initiated by Mauss and Weber and the fragmentary approach to social life (what Durkheim called social fact) as a means to drawing larger inferences (2016, 6–9).
18. By uniformity effect I mean the efforts toward standardization and quantification that seek to flatten localities and other contextual specificities during REDD's implementation.

Chapter 1

1. A synecdoche is a figure of speech in which a whole stands for a part, or a part stands for a whole, as in referring to "wheels" in representation of one's car, or "bread and butter" in reference to one's livelihood.
2. See https://news.delta.com/delta-offsets-carbon-emissions-170000-customers-thursday.
3. Joseph Masco (2010) describes how the nuclear arms race provided a new perspective of the planet as an integrated biosphere and, with it, a new perception of "global risk."
4. Years before, in the 1960s, Rachel Carson's book *Silent Spring* had also contributed to this interconnected way of seeing the world by alerting people to the global effects of pesticides used in agriculture.
5. Geoengineering includes all large-scale interventions in the Earth's natural systems to counteract climate change (http://www.geoengineering.ox.ac.uk/what-is-geoengineering/what-is-geoengineering/).
6. Edwards explains that format means computer processable, space is about uniform grid points, and time is the uniform time steps used by the computer models (2001, 46).
7. According to Edwards, this was a common practice by meteorologists, but by the 1960s it started being automated, a process that required "explicit, computer programmable theories of error, anomaly, and interpolation," rendering the smoothing of the data invisible (2001, 46).
8. These include atmospheric residence time, radiative signature, and photochemical reactivity (Demeritt 2001, 316).
9. GHGs have different warming potentials, depending on their direct and indirect radiative effects and feedback processes.
10. James Carrier and Daniel Miller (1998) noted how society has been increasingly shaped by economic models to the extent that instead of simulating hypothetical future behaviors, these models become prescriptive.
11. See especially Knorr-Cetina 1999, 246, and Zaloom 2006, 170.
12. Edwards concludes that since policy communities depend on these models for advice, "global modeling does not merely represent, but in a social and semiotic sense *constructs*, the global atmosphere" (2001, 64). Emphasized in the original.
13. Fieldnotes, October 16, 2018. This scientist served the US Department of Defense as well as numerous American administrations in public offices. He authored various chapters of the IPCC Assessment Reports.
14. Despite being an important landmark, the creation of the UNEP still reveals the marginal weight of the environment in the international agenda, as the UNEP is a program and not an independent agency inside the UN.
15. The World Meteorological Organization was created following an international conference in Leipzig in 1872 and the First International Congress of Directors of Weather Services, in Vienna in 1873 (Fleming 1998, 42). European countries, Russia, and the United States established their weather services, col-

lecting data systematically and cooperating internationally, in the nineteenth century.
16. According to Gerald Kutney, "climate change was thrust into the political spotlight by three initiatives: the Villach Conference in 1985, the *Brundtland Report* in 1987 and the Toronto Conference in 1988" (2014, 7).
17. Candis Callison refers to vernacularization as the interpretative process "by which a term [like climate change] comes to gain meaning within a group and the work of translation that such term must undergo in order to integrate it into a group's worldview, ideals, goals, perceptions, and motivations to act" (2014, 13).
18. The IPCC also writes methodology reports, special reports, and technical papers (Kutney 2014, 19).
19. For this reason, some countries from the Global South required that an independent body be created to carry out these assessments (O'Riordan et al. 1998, 368).
20. My emphasis.
21. For a detailed description of their roles, responsibilities, and nomination processes, see Kutney 2014, 21.
22. The first, published in 1990, concluded that the world was warming, but could not determine "whether the warming was caused by natural processes or by humanity's greenhouse gas emissions" (Weart 2008, 157). In the second Assessment Report, published in 1998, the anthropogenic origin of climate change was deemed more likely than not (Edwards and Schneider 2001, 221).
23. Emphasized in the original.
24. See Scott (1998, 21) for a similar process in scientific forestry, and LiPuma (2017, 17) for financial markets.
25. This type of policy is one "under which the government commands each firm to produce no more than a certain volume of pollution and specifies the pollution-control technology used" (O'Sullivan and Sheffrin 2006, 202).
26. Simultaneously, European policymakers were pressured by NGOs such as the Environmental Defense Fund to follow the American model created with the Clean Air Act (MacKenzie 2009a, 151).
27. Since the European market is inherently linked to Kyoto's goals, the fungibility rule applies also to all the emission credits from these four markets.
28. Because the United States never ratified the Kyoto Protocol, many of the companies using these markets are American, and most of the offsetting projects involve the forest sector.
29. These countries were the most adamant in the defense of a market mechanism to curb emissions, arguing with the success of their previous experience reducing CFCs. Importantly, they were more interested in emissions trading only, rather than project-based trading mechanisms (like the CDM and the JI) due to their expected higher transaction costs (Yamin 2005, 7).
30. The name "clean development mechanism" embodies this logic of committing poor countries to a "clean," not polluting and therefore sustainable, development.

31. According to the IPCC, the greenhouse gases are carbon dioxide (CO_2), methane (CH_4), nitrous oxides (N_2O), chlorofluorocarbons (CFCs), hydrofluorocarbons (HFCs), hydrochlorofluorocarbons (HCFCs), perfluorocarbons (PFCs), and hexafluoride (SF_6). Of these, HFCs, HCFCs, PFCs, and CFCs are the ones with the highest warming potential.
32. See https://unfccc.int/process/transparency-and-reporting/greenhouse-gas-data/greenhouse-gas-data-unfccc/global-warming-potentials.
33. IPCC's estimates of global warming potential's commensurability are, at times, "admitted to be gross oversimplifications: the effects and lifetimes of different greenhouse gases in different parts of the atmosphere are so complex and multiple that any straightforward equation is impossible" (Lohmann 2008, 361).
34. Elsewhere I argued (Machaqueiro 2017) that these markets actually trade atmospheric space to be filled up with more emissions. Such atmospheric space is created by emissions-reduction projects.
35. See http://www.worldwatch.org/node/5134.
36. Despite the existence of multiple regional carbon markets (significantly smaller than the ones I referred to above), like the Californian cap-and-trade mechanism, the Swiss, or several Chinese regional markets, all of them are based on the premise of a global atmosphere.
37. The work of Sidney Mintz (1985) on sugar provides a good example of this type of commodity, as well as Paige West's (2012) on coffee, Genese Sodikoff's (2012) on timber, or Fernando Coronil's (1997) and Timothy Mitchell's (2011) on oil.
38. The "fat cats" refer to the companies making the biggest profit this way. According to this Sandbag report, the companies were ArcelorMittal, Lafarge, Tata Steel, ThyssenKrupp, Riva Group, Cemex, Holcim, Heidelberg Cement, Italcementi, and Salzgitter.
39. Market advocates recognize that poor countries' authorities and project developers "have an incentive to overstate the amount of emissions reduction achieved," thereby increasing their revenues (Michaelowa 2005, 290). Lohmann also mentions the "perverse incentives for credit seekers" to define baselines lowest as possible and the "business as usual" scenarios the "highest-emitting possible, in order to make the proposed projects appear to be saving as much carbon as possible" (2009, 511–13).
40. For instance, taxes on tobacco prices reflect the costs of treating smoking-related diseases by national health systems. Baron and Colombier explain that in economic modeling, a tax is equivalent to emissions trading in terms of efficiency, that is, both lead to a change in behavior (2005, 156).
41. European carbon permits, traded at about €30 per ton, were hovering at about €5 per ton or less in 2013 (Reed 2013), whereas primary CERs (emissions reductions generated by projects in the CDM), traded at around US$12 in 2007 and 2008, reached US$0.52 in April 2013, registering a fall of 91 percent from a year before (Carr 2013). In 2015, the World Bank recognized that 85 percent of emissions "are priced at less than US$10 per tCO2e" (Kossoy et al. 2015, 17).

42. According to this estimate, the social cost of carbon will be more than US$50 by 2035 and US$71 by 2050 (Trexler 2016).
43. For a detailed explanation of how the social cost of carbon is calculated, see https://www.carbonbrief.org/qa-social-cost-carbon.
44. This commission, chaired by Joseph Stiglitz (Nobel Laureate in economics) and Lord Nicholas Stern, is supported by the Carbon Pricing Leadership Coalition (CPLC), which is a World Bank Group Initiative. This report states that "modeling exercises to calculate the global social costs of carbon have produced numbers that *probably underestimate these costs by very large margins*" (CPLC 2017, 52), emphasis by the report's authors.
45. My emphasis. For the joint statement, see https://icapcarbonaction.com/en/?option=com_attach&task=download&id=505.
46. Carbon prices are affected by patterns in fuel prices because electricity utilities are significant GHG emitters and their choices of energy types to produce electricity can increase or reduce emissions. So, for instance, if natural gas prices are lower in relation to coal, utilities shift to gas, reducing their emissions, and vice versa if prices of coal are lower. In the same way, weather patterns can increase energy consumption, thus increasing emissions.
47. According to Marcu, there was a significant resistance from some parties "to the use of the word 'markets' in the UNFCCC text, as well as any language that could be directly related to markets." To avoid making reference to "credits" and "allowances," the expression ITMO, which was perceived as more neutral, was introduced (2016, 7).
48. This environmental NGO is known for pursuing market-based approaches to address environmental problems. EDF was a key proponent of the sulphur dioxide market and also worked hard to include market mechanisms in Kyoto (MacKenzie 2009a, 151).
49. According to this report, the biggest volume of trading is in the price range of US$1 per tCO_2e. Description of this event is from fieldnotes, October 26, 2016.
50. I am citing here a press article because it provides a clearer explanation of the fund than my own notes of the event.
51. What was not very clear in his presentation was that this "trigger mechanism" is supposed to be paid by taxpayers.

Chapter 2

1. To facilitate the reading, see table 1, identifying all the COPs organized since the beginning of the UNFCCC and concomitant achievements of each one of them.
2. In 1983, the UN secretary general nominated the former Norway prime minister Gro Brundtland to head a commission on environment and development. The commission's final report, *Our Common Future*, is also known as the *Brundtland Report*.
3. The UNFCCC was only one of the five documents that emerged from the Earth Summit. The others were the *Rio Declaration*, the *Convention on Biodiversity*, a *Declaration on the Forest*, and the *Agenda 21*.

4. Canada withdrew from Kyoto before the first commitment period was over and did not agree on a second period because Kyoto did not cover emissions from the United States and China—the two greatest emissions at that point. Belarus, Japan, New Zealand, Russia, and Ukraine declined to sign up for a second agreement period under similar claims.
5. Something that I will explore in chapter 3.
6. Interview, June 25, 2013.
7. According to linguistic anthropologist Kathryn Woolard, language ideology consists of the set of "representations, whether explicit or implicit, that construe the intersection of language and human beings in a social world" (1998, 3).
8. One of my interlocutors inside the NGO Environmental Defense Fund used to work pro bono in an African country's delegation.
9. Povinelli talks about limits on creativity and imagination (2002, 7). Similarly, one could speak of limits on imagining different forms of perceiving the problem of climate change and on creating alternative solutions.
10. https://en.oxforddictionaries.com/. Accessed January 2017.
11. "Fractal recursions involve the projection of an opposition, salient at one level of relationship onto some other level" (Irvine and Gal 2000, 38). Gal states that to be fractal, a distinction must be co-constitutive, so that the terms define each other (2005, 26). Here, I am stretching this idea and considering previous COP decisions as co-constitutive of these ones, not in the sense that their terms define each other, but because the current decisions cannot be fully understood without knowing the terms of the previous ones.
12. According to Gal, "erasures are forms of forgetting, denying, ignoring, or forcibly eliminating those distinctions or social facts that fail to fit the picture of the world presented by an ideology" (2005, 27).
13. The importance of consensus is not just part of diplomatic writing. Within the IPCC, decisions on what to include, or how to draft the scientific report, are made by consensus (see Kutney 2014). According to Büscher (2013), consensus (along with anti-politics and marketing) is a crucial element in the politics of neoliberal conservation.
14. Fieldnotes, December 11, 2015.
15. See http://news.bbc.co.uk/2/hi/8426835.stm.
16. Interview, July 8, 2014.
17. Interview, July 2014.
18. As I have shown in the previous sections, the recursivity and the bureaucratic genre preclude alternative suggestions, perspectives, or narratives to those that are already authorized by the UNFCCC.
19. The use of acronyms is already a linguistic practice of exclusion.
20. Code-switching is the use of more than one language in the course of a single communicative episode. According to linguist Monica Heller, code-switching can best be understood by placing it in the double context of the speech economy of a multilingual community and of the verbal repertoires of individual members of that community (1988, 1).

21. Fieldnotes, November 2013.
22. As I explore in chapter 3, the definition of forests is a highly contentious issue in these discussions.
23. My emphasis.
24. For more information on why defining baselines is such a contentious issue, see chapter 1.
25. See the Annex to Decision 13, "Guidelines and procedures for the technical assessment of submissions from Parties on proposed forest reference emission levels and/or forest reference levels" (UNFCCC 2013a, 36–38).
26. In his analysis of policy documents produced in the context of the meetings at the United States Trade Representative, Dent (2013) argues that policymakers oscillate between specificity and lack of specificity as means to build and reinforce their authority.
27. Fieldnotes, November 2013.
28. The theme of transparency gained its apex in 2009, during the Copenhagen meeting, when suspicions among parties were at its highest, and issues of democracy in the negotiation process were raised more vigorously.
29. Although words can be symbols, I argue that it is the constitution of the acronym that gives these three aggregated words/actions the features of a symbol.
30. The idea of implementing an MRV process over finance was raised by some NGOs and especially by governments of developing countries, who accuse the other group of pledging and not delivering, and of redirecting aid funds for climate change purposes.
31. The *neutral* party is either a team of "technical experts" from both country groups or private companies specialized in carbon accounting.
32. In "Warsaw Outcomes," UNFCCC webpage, http://unfccc.int/key_steps/warsaw_outcomes/items/8006.php, accessed in September 2016.
33. Email communication, March 2014.
34. Interview, February 2014.
35. Interview, December 2013.
36. Fieldnotes, November 2013.
37. This finance platform would imply that, instead of having 195 parties to discuss, there would be only a few, as in the G8 or the G20 models.
38. Fieldnotes, November 2013.
39. The reference to the "ghosts" is inspired by a piece in the *Guardian* (Goldenberg and Vidal 2010a).
40. See Vidal and Goldenberg 2014.
41. See Carrington 2010.
42. See Goldenberg and Vidal 2010b.
43. It is important to make a distinction between the large US-based NGOs and the smaller ones, which are very critical of the results achieved in Warsaw. The large US-based NGOs might express some frustration over the achievements, but do not classify the COP as a failure.

44. According to the Amazon Fund website, the government of Norway provided more than US$646 million during 2013. See http://www.amazonfund.gov.br/FundoAmazonia/fam/site_en/Esquerdo/doacoes/.
45. Namely, the national monitoring systems, or even the information hub decided during Warsaw.

Chapter 3

1. The coalition covers the three largest tropical forest areas (Amazonia, Congo Basin, and New Guinea), including other nations with rainforest resources. Tellingly, CfRN's website defines coalition members not as countries with tropical rainforest, but as "developing nations with tropical rainforest resources"—a utilitarian language that sees forests merely as an economic resource to be managed (Scott 1998, 13). See http://www.rainforestcoalition.org.
2. These percentages vary according to different sources. This one is from The REDD Desk, a collaborative platform that gathers information about REDD. See http://theredddesk.org/resources/why-redd-matters.
3. The Nature Conservancy (TNC) is, like the Environmental Defense Fund (EDF), an NGO that promotes market-based solutions for environmental problems. They are both part of what some authors call the "big green." See Klein (2014) for detailed descriptions of the modus operandi of these "big green" organizations.
4. Under the CDM, countries can submit afforestation and reforestation projects originating temporary certified emission reductions (tCERs) to be traded in the market. See chapter 1.
5. Referred to as nonpermanence, it "refers to the temporary nature of [carbon] removals, given that carbon contained in the biomass of trees is at continuous risk of being emitted into the atmosphere" by fire or deforestation. See http://unfccc.int/land_use_and_climate_change/lulucf/items/3064.php.
6. Referred to as leakage. It is defined as "those emissions of greenhouse gases outside the area of the project as a result of its implementation." See http://unfccc.int/land_use_and_climate_change/lulucf/items/3064.php.
7. According to the CDM pipeline database, afforestation and reforestation projects represented 0.8 percent of the total amount of approved projects (see http://www.cdmpipeline.org/cdm-projects-type.htm). Because industrialized countries would be able to meet all their reducing emissions targets using forest carbon sequestration projects (something unacceptable for poor countries) the Kyoto Protocol established provisions limiting the amount of such projects used by industrialized countries.
8. See Yamin and Depledge 2004, 181–82.
9. My emphasis.
10. Interview, May 21, 2013.
11. I was referring to the UNFCCC's stated goal of having a new legally binding agreement to replace Kyoto, with an operating REDD mechanism by 2020.
12. Interview, May 30, 2013.

13. See https://climatenewsnetwork.net/forest-carbon-storage-puzzles-scientists/.
14. The uncertainties about forest carbon are discernible in discourses about tropical forests: while much of this literature claims that science has proven that "primary" mature tropical forests are the most efficient carbon sinks, more recent studies have shown that "secondary" forests can actually absorb more carbon during their growth. For more on this, see McSweeney 2014.
15. Interview, May 28, 2013.
16. See http://www.un-redd.org.
17. See https://www.forestcarbonpartnership.org.
18. The Amazon Fund was also created in 2008.
19. Interview, June 24, 2013.
20. See http://www.forestcarbonportal.com/content/redd-finance-whos-counting.
21. Interview, June 24, 2013.
22. Acronym for Official Development Assistance, a term coined by the Organization for Economic Co-operation and Development (OECD) to measure aid.
23. Interview, June 24, 2013. In translating this conversation from the Portuguese, I have maintained the gendered references to countries. I have also italicized the words that were said in English during the conversation held in Portuguese.
24. At the time of our conversation, the Norwegian government had changed recently, and my interlocutor emphasized that the new government maintained the same position on forest conservation and REDD issues.
25. Interview, July 17, 2014.
26. Interview, August 21, 2014.
27. My emphasis.
28. My emphasis.
29. The process followed by Acre's government is extraneous to the rules of the Amazon Fund, and in fact, this state policy created frictions between state and federal governments. For more on this, see chapters 4 and 5.
30. See http://wrm.org.uy/other-relevant-information/how-does-the-fao-forest-definition-harm-people-and-forests-an-open-letter-to-the-fao/.
31. That is because restoration or reforestation "does not increase forest area, as it occurs on lands already defined as forest" (Chazdon et al. 2016, 544).
32. The Chinese demand for timber, and how it has promoted illegal logging and corruption, is frequently mentioned in international reports, and the Mozambican media. See MacKenzie (2006), Environmental Investigation Agency (2013), Ekman, Wenbin, and Langa (2013). Also, *Canal de Moçambique* (Mulungo 2016, March 23, 2016), and *Savana* (March 25, 2016). During an event with Mozambican authorities, World Bank officials, and members of NGOs and local organizations in Zambézia, the problem of illegal logging in the Gilé National Park was openly discussed. Many mentioned the fact that government officials were themselves involved in that illegal activity, while the administrator of the park went even further, denouncing the presence in that same room of people involved in such environmental crime (Fieldnotes, April 20–21, 2016).

33. A safeguard is "a policy, procedure or process that accompanies an investment, project, program, policy or other activity and is designed to mitigate social and environmental risks" (Daviet et al. 2013, 2).
34. Such as the Tropical Forestry Action Plan (see Colchester and Lohmann 1990; Halpin 1990) or the Norwegian carbon plantations in Tanzania and Uganda (cited in Fogel 2005).
35. See UNFCCC Decision 1/CP. 16, appendix I, paragraph 2, 26–27 for a complete description of safeguards.
36. Cris Shore argues that policies function as "political technologies," disguising their political nature by objective, neutral, and legal-rational idioms (2011, 171). These political technologies can be understood as governmentality techniques.
37. Both expressions are used interchangeably; however, I was told by an environmental NGO practitioner that "non-carbon benefits" is a better expression because the prefix "co" implies that those benefits are secondary to carbon emission reductions, and the idea was to give them the same importance as carbon emission reductions (Interview, May 13, 2013).
38. My emphasis.
39. Fieldnotes, November 21, 2014.
40. The Rio Branco Declaration was released on August 11, 2014, and states the commitment by the Governors' Climate and Forests Task Force (GCF) to reduce tropical deforestation in their jurisdictions by 80 percent by 2020 if adequate results-based funding is made available. The GCF is a subnational collaboration between twenty-six states and provinces in Brazil, Indonesia, Mexico, Peru, Spain, and the United States. See http://www.gcftaskforce.org/documents/2014_annual_meeting/GCF_RioBrancoDeclaration_August_5_2014_EN.pdf.
41. See http://www.bionic-planet.com/2017/11/15/what-can-an-obscure-trademark-dispute-tell-us-about-global-climate-talks-plenty/.

Chapter 4

1. Hoelle also refers to these tropes used by urban Brazilians to describe Acre, adding "where the wind turns around" and "where Judas lost his boots" (2015, 1).
2. Acre was formally incorporated in Brazil in 1904 after a conflict with Bolivia.
3. In 2010 the state of Acre signed a Memorandum of Understanding with the states of California and Chiapas in which the state of California committed itself to buying carbon offsets generated by the forests of Acre and Chiapas. This agreement has been contested in the three locations. See https://www.arb.ca.gov/lispub/comm/bccommlog.php?listname=tfs2018.
4. In Portuguese, the acronym stands for Sistema de Incentivos a Serviços Ambientais.
5. The VCS was founded in 2005 by the Climate Group, the International Emissions Trading Association, the World Economic Forum, and the World Business Council for Sustainable Development. The CCB Standards were created by the Climate, Community and Biodiversity Alliance (CCBA), a partnership of several

NGOs (CARE, Conservation International, The Nature Conservancy, Rainforest Alliance, and the Wildlife Conservation Society) created in 2003 with the purpose of developing standards for the implementation of environmental and social safeguards. In November 2014, VCS began to directly manage the CCB Standards.

6. For a detailed explanation on what jurisdictional REDD is, see Fishbein and Lee 2015.
7. It is a situation in which deforestation moves from one area where REDD is being implemented into a new one where deforestation was not occurring.
8. Despite its catholic matrix, CIMI does not evangelize Indians. Instead, it works closely with indigenous communities and often represents their interests within the federal government and other organizations. Inspired by liberation theology, CIMI has a long history of struggle for the rights of Indians and poor rural populations. For that reason, CIMI is frequently treated with hostility by authorities.
9. Namely reports and policy briefs prepared by NGOs referring to Acre's experience in REDD and documents prepared by Acre's authorities.
10. The idea of financially compensating those who preserve an ecosystem follows from the premise that, to preserve nature, it is necessary to give it a price. REDD is a form of payment for ecosystem services. Looking at the environment in terms of economic benefits (conservation) and costs or losses (environmental deterioration) was mainstreamed in the sequence of a study commissioned by the environment ministers from the G8+5 and published in 2007. See www.teebweb.org.
11. Although the official myth did not disclose the identity of the NGOs promoting the REDD discussions and workshops, I later found that at least the US-based organizations Forest Trends and WWF were particularly active in this endeavor.
12. Later in my fieldwork, Márcia left the governmental office to become a consultant inside an American environmental NGO working with several Amazonian states, Peru, and Bolivia.
13. Interview, July 16, 2013.
14. Interview, September 22, 2015.
15. The Portuguese name he used was "dinamizadores de serviços ambientais." Interview, October 9, 2015.
16. See Cabello and Gilbertson (2011) for an example of these critiques.
17. The "No-REDD" group is a coalition of NGOs, activists, indigenous organizations, and academics that have consistently denounced REDD and worked against the establishment of a forest-carbon market. Both in Acre and in Mozambique I found authorities referring to this international group in derogatory terms. See Cabello and Gilbertson (2011) for an overview of this coalition.
18. The Pilot Program to conserve the Brazilian Rainforest (PPG7) was launched in 1992 to reduce deforestation in the Amazon. The program was financed by the G7 (Canada, United States, France, Germany, Italy, United Kingdom, and Japan) and the European Union.

19. Some of those who oppose REDD accuse Steve Schwartzman, an important member of EDF, and a "star" in the world of environmental activism, of using his "alleged friendship" with Chico Mendes to claim that REDD and other forms of ecosystem payments would be supported by Chico. Schwartzman was indeed one of Chico Mendes's contacts in the United States and helped him reach international audiences. Fieldnotes, June 29, 2013.
20. WWF has an important presence in both Acre and Mozambique, with local offices in Rio Branco and Maputo, and works closely with local authorities in the implementation of environmental programs, including REDD.
21. Although not present in Acre, IUCN has offices in Maputo and has been part of the REDD process in both locations.
22. In 2007, state authorities carried out a detailed economic and ecological zoning plan enabling the establishment of forest management activities and other initiatives to explore the state's resources and allowing Acre's authorities to *see like a state* (Scott 1998).
23. It is not surprising that this American scientist is perceived by some of my interlocutors as a CIA agent, since many American citizens working or doing research in the Amazon are typically indexed as such. However, it is significant that American citizens continue to be associated with what is understood as a challenge to Brazil's sovereignty over Amazonia. Also significant is that in the absence of knowledge about the work developed by this scientist, such a knowledge gap would be explained by the secrecy that the work as a secret agent would entail.
24. See Jakobson (1960, 1995) and Waugh (1980) for poetic function and other communicative functions.
25. The term civil society acquired great currency in the late 1970s, especially after state socialism projects in Eastern Europe started to collapse and the separation between state and society became apparent (Young 1994, 34–37). However, the idea of civil society "has proven impossibly difficult to pin down. The more its advocates have sought to make it a mantra of sociomoral regeneration and social analysis, the more elusive and ambiguous it has become" (Comaroff and Comaroff 1999, 5).
26. At the local level, fifteen NGOs and eight scientists; at the national level, fourteen NGOs, eight state governments, two international cooperation agencies, five scientists, and two companies; at the international level, fifteen NGOs, eight governments, ten researchers, and ten companies from the carbon market sector (Governo do Estado Acre 2012, 18).
27. Embrapa, the Brazilian Agricultural Research Corporation, is a state-owned company and highly involved in cooperation initiatives in poorer countries in the agricultural sector, like the Prosavana project in Mozambique.
28. CCBA, as aforementioned, is one of the leading certifying agencies for carbon credits.
29. Amazon Environmental Research Institute (in Portuguese, Instituto de Pesquisa Ambiental da Amazônia) is a national NGO involved in providing recommen-

dations for public policies on the environment. Its board includes American scientists Steve Schwartzman and Daniel Nepstad, both with a long history of research in Amazonia.

30. Getúlio Vargas Foundation (in Portuguese, Fundação Getúlio Vargas)—based in Rio de Janeiro and named after a very popular Brazilian president—is an institution for graduate studies and policy research, known for its neoliberal bend.
31. WWF, IUCN, and EDF were already involved in prior initiatives of forest conservation in the Amazon.
32. Interview, July 17, 2013.
33. At the time of this interview, the minimum wage in Brazil was R$678 (around US$310 at that time).
34. Interview, July 8, 2013.
35. The Green Grant (Bolsa Verde) is a cash transfer program created by federal Law 12.512 of October 14, 2011. It provides R$300 every three months for families living in environmental conservation areas, as long as they respect the rules of sustainability in the use of environmental resources. For more information on the program, see http://www.mma.gov.br/desenvolvimento-rural/bolsa-verde.
36. Her use of the word "alienation" points to a Marxist meaning of lack of political consciousness.
37. The acronym stands for Gesellschaft für Internationale Zusammenarbeit GmbH.
38. The acronym stands for Kreditanstalt für Wiederaufbau.
39. The SISA law has several donors: initially, the state received R$240,000 from the German Cooperation Agency (GIZ), WWF-Brasil, and International Union for the Conservation of Nature (IUCN) (WWF-Brasil 2013, 43); on a second phase, WWF and the Amazon Fund funded another part of SISA's institutional development. "Additionally, Acre signed two payment-for-performance agreements with the German Development Bank (KfW) in 2012 (four-year payment period) and 2013 (single payment) through the bank's Global REDD Programme for Early Movers" (Duchelle et al. 2014, 36).
40. Roberto compared the case of Acre with the state of Rondônia, which continues to base its development in cattle ranching and logging. Interview, September 21, 2015.
41. CEVA integrates four civil society organizations and four governmental entities. Supposedly, it is through CEVA that civil society participates in the system.
42. CDSA is supposed to assist in financing projects to develop environmental services and trade carbon offsets.
43. In 2018 state authorities created a new website about the SISA law containing information about its governing structure and the activities included in the legal framework. I was told that the website was a response to German donors demanding more transparency toward the public.
44. Dent describes a similar process through which multinational corporations support NGOs so they can act locally in the pursuit of the interests of these corporations. While these "practices appear to be uniform, from Boston to Bei-

jing, Buenos Aires to Bali," a localized approach is needed to understand these processes (2016, 425).
45. Fieldnotes, August 26, 2015.
46. See http://www.teebweb.org.
47. Interview, July 16, 2013.
48. He was referring to natural rubber, from the tree *Hevea brasiliensis*.
49. Interview, July 17, 2013.
50. For my interlocutor at the ministry, the crucial piece of legislation is the Forestry Code, which then can be complemented by state codes. According to her, the idea that Brazil needs a federal legislation connecting state programs (as claimed by NGOs) follows from demands by foreign investors, who want legal guarantees and reduced financial risks (Fieldnotes, June 27, 2013).
51. Chapter X of the Law 12.651 of May 25, 2012 (also referred to as the Forestry Code).
52. The publication did not specify which states ordered this legal opinion, but the bibliography refers to a report prepared by Ludovino Lopes Attorneys in 2012, São Paulo, for IDESAM (Institute for the Conservation and Sustainable Development of the Amazon) and the Governor's Climate and Forest Task Force (GCF). Later, I would find out that IDESAM had been hired by the World Bank to work as consultants in the implementation of REDD in Zambézia, Mozambique.
53. Prior to the meeting, parties submitted their estimates for emissions reductions, referred to as "contributions."
54. Envirotrade was founded by Robin Birley (England) and Philip Powell (South Africa) in 2002. In 2003, the company was awarded a €1.5 million grant to develop a carbon offset project with the University of Edinburgh, in N'hambita (see http://envirotrade.net). The project claimed to be developing agroforestry systems, protecting biodiversity, and generating carbon finance income to provide sustainable local livelihoods, and has been certified by the Rainforest Alliance, Plan Vivo, and the CCBA. However, local and international activists challenged these claims, accusing the project managers of not paying the communities the incomes promised, of occupying fertile soil with tree plantations instead of food crops, and of increasing gender and economic inequalities. Due to these critiques, the European Commission suspended its last €450,000 payment for the project. For more on the N'hambita project, see Kill 2013.
55. In a slightly different version of this story, the customs officer would refer instead to the Naamacha border toward Swaziland.
56. By mythological features I mean its composition of different elements (which produce meaning through the combination of their relations), the repetition of its telling, and the fact that each time, the story presented slightly different variations (Lévi-Strauss 1955). All these different variations produced the same meaning that I explore in the chapter.
57. I say *then* because after the 2014 general elections these two ministries suffered a profound reorganization. MINAG became the Ministry for Agriculture and

Food Security (MASA), and MICOA became the Ministry for Land, Environment and Rural Development (MITADER). I keep all acronyms in their Portuguese format.
58. At the time of these interviews, Mozambique was under a "non-declared war" in the rural areas, spurring fears of a new civil conflict triggered by the election's outcome. General elections were held in October 2014. See Cordeiro 2013.
59. Fieldnotes, July 8, 2014.
60. This was a joint interview on July 7, 2014, with two members of the NGO. Most of it was conducted in English, because one of the staffers did not speak Portuguese well enough, but at times, the other person shifted to Portuguese, since her English level was also limited.
61. The history of MICOA's creation, back in 1988/1989, contradicts this vision of a purposefully weak ministry. The idea behind "coordination" was based on the assumption that the environment is a transversal issue and therefore needed to be involved in all the other ministries. The person who told me about the history of MICOA was, at that time, actively involved in its creation. When we talked, however, he recognized MICOA's incapacity to actually regulate environmental issues (Fieldnotes, July 15, 2014). A governmental officer confirmed that when MICOA was created, it was actually very strong and was responsible for very important legislation. At the time of our conversation in 2014, though, he agreed that MICOA had become weak, a fact demonstrated by the choice of the minister (Ana Paulo Chichava), as well as some of the people leading the several directorates (Fieldnotes, July 4, 2014).
62. At the time of these discoveries, Mozambique registered one of the highest GDP growth rates (between 7 and 8 percent). See Smith 2012.
63. Interview, Maputo, July 8, 2014.
64. Interview, April 1, 2016.
65. This information about the cooperation with JICA comes from a printed presentation, prepared by the Ministry of Agriculture for the IX National Meeting of Lands and Forests, Nampula, May 22–23, 2014. I did not attend this meeting, but some of its materials were given to me by some of my interlocutors during fieldwork.
66. See chapter 1.
67. Green Resources began planting forests in Mozambique in 2007 and manages several thousand hectares of forest in Tanzania and Uganda (www.greenresources.no). According to the company's website, Green Resources is also one of the first companies globally to receive forest carbon revenue through reduction emissions for the CDM. In 2009 the company established an agreement with the Mozambican government to develop 126,000 hectares of forest plantation over a period of fifteen years in Niassa and Nampula. These plantations have been strongly contested due to several grievances with local communities who complain about not being consulted in the process of land alienation, having trouble accessing natural resources, and not being offered jobs as promised by

Green Resources. As a result, some of these plantations have been burned down by the communities.
68. According to article 33, a project proponent must pay 100,000 MT to submit the project, 100 MT per hectare after project approval and before the license is issued, 10 MT per hectare as annual fee, and 40 MT per hectare upon the license renewal. The proponent is also mandated to pay 10 percent of each ton of carbon credits traded in either voluntary or compliance markets. The only exception to these payments is when local communities are themselves project proponents (República de Moçambique 2013, 1072).
69. This 20 percent follows the same logic of the 20 percent that logging companies must provide to local communities. During a conversation with a legal consultant working for the government, I was told that logging companies pay these fees to the communities; the problem is that they pay 20 percent over the timber they declare and not over what they actually cut down and sell. The same could happen with a REDD project.
70. Interview, Maputo, July 7, 2014.
71. Fieldnotes, July 4, 2014.
72. In Portuguese, *direito ao uso e aproveitamento da terra*. This right applies to individuals and communities through customary law and specific requests to authorities, and it also applies to national citizens occupying the land for more than ten years (Serra 2013, 60).
73. This explanation was given to me by a current member of the Environment Ministry but who was, at that time, working as a consultant on the behalf of a local NGO. In this role, he participated in the discussions leading to the drafting of the REDD law (Fieldnotes, June 30, 2014).
74. Natércia was part of the Ministry of Agriculture, and after the 2014 changes, she shifted to the Ministry of the Environment. Interview, July 10, 2014.
75. Interview, March 18, 2016.
76. Interview, March 29, 2016.
77. In 2013, while in Maputo I was able to contact a Mozambican national working at the bank, who promised me an interview. As is frequent in Mozambique, this interview was scheduled by text message. Two days prior to the interview, he asked me to send him my questions by email. I drafted a few very general questions about the REDD process, hoping to get into the details when talking to him in person. On the day of the interview, he called me to change the time of our meeting to late that night because he wanted the focal point in D.C. to join our conversation by Skype—he "did not feel comfortable" answering those questions by himself, he told me. I protested, explaining that I could not walk half of the city by myself, in the dark, and that given the short notice I would not be able to get a ride or someone to go with me. We decided to cancel the interview under the agreement that this D.C. focal point would receive me later, upon my return to the United States.

78. These two companies established a consortium because only one had an office in Maputo. However, this Portuguese company with a Maputo office is an engineering and construction company, not a consultant.
79. It seems that the R-PIN was prepared by the same Brazilian NGO that was later commissioned by the World Bank to prepare other documents.
80. While the R-PIN serves as an expression of interest for FCPF's support, the R-PP is supposed to provide a more detailed description of the institutional arrangement of the country to address deforestation.
81. FAS was created in 2007 in the Brazilian state of Amazonas. It is financed by the Brazilian bank Bradesco, Coca-Cola Brazil, the Amazon Fund, and Samsung.
82. Jerry Dávila (2010) also explores this similarity argument in Brazilian diplomacy in Africa, calling it "tropical technology," that is, forms of technology that, being appropriate to the tropical hurdles of Brazilian reality, could also serve Africa's needs. See pages 223–28.
83. For those involved in this "South-South Cooperation," the fact that Brazilian companies and authorities were acting as neocolonial parties in other sectors of Mozambique (notably in mining and agriculture) was never an issue.
84. Fieldnotes, March 17, 2016.
85. I was never able to figure out the reasons for the end of this cooperation, or even whether the cooperation was a one-time experiment or if it was intended to be a long-term relationship.
86. Bolsa Família is a conditional cash-transfer program introduced by the socialist government of President Lula da Silva in 2003, as part of wider federal policies to fight poverty in the country. Conditions to receive the money include keeping the children in school and following the vaccination program. See http://www.caixa.gov.br/programas-sociais/bolsa-familia/Paginas/default.aspx. Under Bolsa Floresta, families commit to not cutting trees on their properties and are compensated accordingly.
87. I explore this issue in chapter 5. As mentioned before, although I could not confirm this, it seems that it was Celso's organization that wrote Mozambique's R-PIN.
88. This discrepancy follows from the different definitions of forest that the documents adopt. The old one states that a forest is a 1 ha area with 10 percent or more forest cover of trees higher than 5 meters (World Bank and República de Moçambique n.d., x); the new one adopts the current definition suggested by FAO—a 1 ha area with 30 percent or more forest cover of trees higher than 5 meters (República de Moçambique 2016, 48).
89. It is also possible to notice this type of corrections just by examining the R-PIN and the World Bank review of it. See https://www.forestcarbonpartnership.org/mozambique.
90. Interview, Maputo, July 17, 2014.
91. That is one of the reasons why international NGOs like WWF invest considerably in local NGOs, leading activists in Acre, or members from JA! in Mozambique to accuse their local counterparts of "selling out."

Chapter 5

1. Interview with Roberto, state officer involved in the creation of the SISA law, September 21, 2015.
2. See chapter 4.
3. Known in Portuguese as *"projetos de assentamento,"* these are defined by the federal government and divided in small agricultural plots to be distributed to poor families. Once allocated to a family, the plot cannot be sold or rented.
4. In a conversation held in Portuguese, Carlos (a member of the local office of a transnational environmental organization) used the expression "tipping point" in English. We talked in a meeting room that had a white board filled with scribblings and diagrams related to this NGO's activities in Acre. Many of the words were in English, which can be explained by the fact that this NGO is based in the United States, but it also points to the fact that "development" is dominated by the English language (Fieldnotes, October 8, 2015).
5. See chapter 4.
6. See http://www.v-c-s.org/project/jurisdictional-and-nested-redd-framework/; accessed December 13, 2017.
7. Squatters (in Portuguese, *posseiros*) are very common in these very large properties and often have usucapion rights over the lands they occupy.
8. Interview, Rio Branco, July 16, 2013. As far as I could tell, the examples given by Márcia were based on concrete situations. Throughout my fieldwork I was told several times by activists (and one state official) that the ownership of the land where one REDD project was being developed was legally dubious and being contested.
9. There are five private projects, all of them developed by the same US company, although belonging to different landowners. Two landowners, however, are connected by family ties: two of the projects are owned by the brother-in-law of another project's landowner.
10. Interview, September 22, 2015.
11. The United Nations Conference on Sustainable Development Rio+20 was celebrated twenty years after the Earth Summit that took place in Rio de Janeiro, Brazil, in 1992. Given the media coverage of the Rio+20, many activists took the opportunity to contest REDD and other market-based conservation policies. Some people in Acre participated in a protest during a special event where Acrean authorities were showcasing the SISA law.
12. Fieldnotes, November 20, 2015.
13. Fieldnotes, October 5, 2015. Blaming Acre's authorities for their choice of the nesting methodology is odd, since the methodology he claimed authorities should have chosen would have made the projects even more imbricated in the state's program. That is, while the methodology chosen by the authorities allows projects to credit their offsets independently of the state, the methodology he mentioned would incorporate the projects in the jurisdiction, making authorities responsible for paying the projects for the offsets generated.

14. For example, in one document detailing the JNR methodologies, in a short section dedicated to "safeguards" it is stated that "jurisdictional programs shall comply with all UNFCCC decisions on safeguards for REDD+ and any relevant jurisdictional (national and subnational) REDD+ safeguards requirement. The jurisdictional program (or baseline) description shall describe how the program meets these requirements" (VCS 2017, 17). There is no information regarding procedures in case the program later fails to comply with such requirements.

    Another part states: "Jurisdictions following Scenario 2 or 3 shall develop a mechanism for receiving, screening, addressing, monitoring and reporting feedback on grievances and concerns submitted by affected stakeholders" (VCS 2017, 18). But nothing is said about what to do with such grievances, and whether jurisdictional authorities can act upon project developers when people express any grievances toward projects.

    It is important to mention that certification procedures rely on evidence provided by authorities and project developers on how the REDD project/program impacts populations (either positively or negatively). Moreover, auditors are taken to the sites and paid by project/program proponents.
15. Although the VCS is based in Washington, D.C., the officer I talked to is placed in São Paulo.
16. I heard from authorities several explanations for not being able to do anything, from lack of resources for visiting the communities affected by the projects to the legal impossibility of acting inside private property. Notwithstanding the reasons invoked by authorities, the fact that REDD projects are sanctioned by international entities (some of them financing Acre, like the World Bank), might have also been a compelling argument to not act.
17. These reserves were created in the context of the intense struggles between rubber tappers and large landowners who intended to clear the forest for cattle ranching (see box 1).
18. Interview, November 12, 2015.
19. Interview, November 12, 2015.
20. See http://www.altinomachado.com.br/2008/12/o-extrativismo-est-falido.html.
21. Interview with Roberto, September 21, 2015.
22. Interview, September 21, 2015.
23. Citing Brazilian national statistics (IBGE), Hoelle states that from 1998 to 2008, the count of cattle in Acre increased by over 400 percent (2015, 3).
24. See chapter 1.
25. The Yawanawá are an indigenous group from the Pano linguistic family. Yawanawá groups can be found in Acre, Peru, and Bolivia. See https://pib.socioambiental.org/en/povo/yawanawa.
26. Interview, September 23, 2015.
27. In Portuguese, Planos de Gestão Territorial e Ambiental. These plans were conceived in consultation with indigenous groups in the context of the state's zoning that took place in 1999. The indigenous plans started in 2004 and are supposed

to help indigenous people to sustainably manage the natural resources of their territories.
28. See http://imc.ac.gov.br/programa-para-pioneiros-em-redd-rem/.
29. Interview, September 24, 2015.
30. KfW funds only paid agents until 2016, and they wanted the government's guarantee that these agents would continue to be paid and that new agents would be trained and certified.
31. The ways in which I refer to indigenous peoples are necessarily too broad and cannot possibly reflect the diversity of opinions between groups and between individuals inside groups. Therefore, my work only traces the perspectives of indigenous leaders with a public profile, and notoriously involved in REDD, either favoring or opposing it.
32. Ninawa is the only person who, after reading my IRB consent form, specifically asked me not to anonymize him in my writing.

    Also known as Kaxinawá, the Huni Kuin are part of the Pano linguistic family and can be found in Acre and Peru.
33. My emphasis. Interview, July 6, 2013.
34. I believe he was referring to McKinsey and Company, an American consulting company that has published numerous studies about climate mitigation. In 2007 McKinsey published a report on the "Abatement Cost Curve" of emissions reductions that is often cited in the policymaking world to justify mitigation options, including REDD. That report was reviewed in 2009, and again in 2010. For more information, see https://www.mckinsey.com/business-functions/sustainability-and-resource-productivity/our-insights/greenhouse-gas-abatement-cost-curves. For a rebuttal of McKinsey's cost curves, see https://www.illegal-logging.info/sites/default/files/uploads/McReddEnglish.pdf.
35. FUNAI, the National Indian Foundation (Fundação Nacional do Índio), is the federal agency responsible for defining and implementing policies related to indigenous peoples and their protection.
36. Interview, October 7, 2015. The name of the building that houses Brazil's Ministry of Foreign Affairs.
37. In February 2018, the German donor (KfW) defined that threshold at 330 square kilometers per year, meaning that if Acre's deforestation rate reached 331 square kilometers in one year, there would be no financial compensation.
38. Statement by Márcia during one of our conversations. Interview, September 28, 2015.
39. Fieldnotes, October 26, 2016.
40. According to one of my sources inside the government, the Germans recommended the creation of a website with more information about the SISA law. This website was only uploaded in the later period of my fieldwork. See http://imc.ac.gov.br.

    This lack of knowledge of the SISA law was easy to assess during conversations I had with rubber tappers in the forest but was also confirmed by other people. For instance, Carlos, a member of an environmental NGO, deplored the wasted opportunity to "teach" people about deforestation by not telling them why they were being subsidized.

41. This expression was used by some interlocutors as a critique to a common situation in which a group of people override entire sections inside the ministries. The UT-REDD is such an example since it is not integrated in any directorate nor is it overviewed by any other than the minister himself, and it is entirely paid by the World Bank. This elitism was equally embodied in the space occupied by the UT-REDD: while the offices of other ministries were situated inside very old buildings either in the center of the city (previously occupied by white Portuguese settlers) or in its outskirts, the UT-REDD rented a house with a garden in the street parallel to the main avenue where most embassies, international organizations (like the bank), and NGOs are located.
42. My emphasis. Interview, February 23, 2016.
43. Interview, March 18, 2016.
44. Arnaldo was referring to the Emission Reduction Purchase Agreement in which the bank defines the price and the number of tons it is willing to buy.
45. Under the ERPA, the bank committed to buy up to 8,724,732 tCO2e, which amounts to US$5.70 a ton. But, more importantly, the letter fulfills the goal of ensuring exclusivity to buy the reductions for a period of five years.
46. Interview, March 29, 2016. Arnaldo used the expression "willingness to pay" in English, amid a conversation held in Portuguese.
47. Interview, March 29, 2016.
48. See https://www.forestcarbonpartnership.org/mozambique.
49. The landscape approach was planned for two jurisdictions, Zambézia and Cabo Delgado, but I did not conduct any research in Cabo Delgado.
50. The REDD National Strategy is also part of the UNFCCC regulation to implement REDD. For more details on how this strategy was prepared, see chapter 4.
51. See http://documents.worldbank.org/curated/en/921041499661351446/pdf/117318-WP-P160033-PUBLIC-Mozambique.pdf.
52. This one was the fourth.
53. While I was in Maputo, I would frequently watch evening news reports on TV about illegal timber from Gilé seized by authorities. This activity is known to involve Mozambican officials, bribed by individuals working for Chinese businesses. In some of those reports, "local communities" complained of not being able to stop the individuals carrying the timber out of the park for fear of retaliation by local authorities. See Mulungo (2016); Senda (2016).
54. This type of tree, *Swartzia madagascariensis*, is similar to rosewood. In 2015, the government expanded an existent export ban, making it illegal to export all raw timber logs of *pau ferro*.
55. See chapter 3 for the different definitions of forest and their implications.
56. Interview, March 29, 2016. This Malthusian anxiety is also very common among environmentalists.

57. Interview, March 18, 2016.
58. In the 1960s and 1970s Mozambique was a world producer and exporter of cashews; however, the civil war greatly weakened cashew production. In the early 1990s the World Bank developed a set of policies with the goal of repositioning Mozambique as a world producer and exporter. In the end, the sector was strongly impacted by international price shocks—especially the small farmers who were not supported by these policies—leading to a major collapse of production and processing facilities. For more on this, see Penvenne 2015, especially the Epilogue, and the World Bank's "revisit" to its own policies in Aksoy and Yagci 2012.
59. Charcoal is produced by keeping piles of wood under a very slow fire. Given the availability of wood and the low technology required to make it, charcoal is an important source of income for many families.
60. While the REDD Strategy cites a study to make an argument about how the high demand for charcoal in Maputo is deforesting Gaza and Inhambane, the strategy fails to cite the same study about the lack of evidence on a direct connection between charcoal production and deforestation:

> The cause-effect mechanism that leads to the loss of forest area is complex and rarely induced by a single cause. It's rather easy to see a relation between unsustainable fuelwood and charcoal production and forest degradation but, in spite of common perception, it's not easy to prove a direct relation between woodfuel production and deforestation. On one hand, it's indubitable that intense and unsustainable charcoal making can lead to heavy degradation or even complete loss of forest cover but, on the other hand, it's also true that fuelwood and charcoal production are very often simple byproducts of the expansion of cropland and shifting cultivation into previous forested lands. In the second case, for instance, a reduction of fuelwood and charcoal demand would do little to reduce the deforestation rate. (Drigo et al. 2008, 34–35)

61. The key word here is eucalyptus (*Eucalyptus genus*). This species is not native to Mozambique and is mostly used by the paper industry for its high cellulose content and rapid growth rate. The most important feature of the landscape program is the large-scale planting of eucalyptus by a Portuguese paper company.
62. Crewe and Harrison state that programs to promote the use of cookstoves were responsible for women's workload increase, while men retained control over the production process and the income generated by stove-making (1998, 12–13).
63. Interview, March 18, 2016.
64. Interview, February 23, 2016.
65. During an interview with a member of the UT-REDD I was told that, by including the conservation parks in the landscape program, they were able to access a different pool of funds (allocated to conservation areas). It is possible that the inclusion of these areas in the program was part of a strategy to access more funds.

66. The call for this alliance was one of the messages of the Brundtland Report (1987).
67. Interview, March 25, 2016.
68. Although a member of the current government, this officer is very critical of the previous Mozambican administrations, of which he was not part. I do not describe his position to protect his identity. Interview, June 30, 2014.
69. See chapter 4.
70. While the first part of this draft had many gaps and blank spaces to be filled after the release of the National Strategy on REDD+, the part about the FIP and the plantation of eucalyptus was complete.
71. The investment by this company is supported by the World Bank's financial arm, the International Finance Corporation. The IFC will make an equity investment equivalent to 20 percent of the initial pilot stage value in the DUAT concessions (República de Moçambique and FCPF 2015, 67).
72. This "challenge" refers to events in which other companies saw their plantations and nurseries burned down by local communities affected by the loss of their lands and unfulfilled promises of jobs.
73. This means that a household that allows the company to plant eucalyptus on its land becomes responsible for that plantation, receiving a payment in exchange.
74. Except for the grievance mechanism and school attendance, all the activities included in this plan for social development are the same ones included in the landscape program: conservation agriculture, training, production of charcoal from eucalyptus, and sustainable forest management.
75. Matos explained that although Mozambique's environmental laws are laxer than in the European Union, his company would still treat the effluents complying with European legislation. Fieldnotes, March 2, 2016.
76. Every time the debate got more tense, the mediators would deflect all disagreements.
77. See Hughes 1998 and Chitará 2003.
78. Fieldnotes, April 20–21, 2016.
79. This empirical explanation reproduces the assumption about the equivalence between biomass and carbon, which is translated into the use of the allometric equation to estimate both. The problem with this assumption is an issue that I explore in chapter 3. See also Lovell and MacKenzie 2015.
80. Interview, February 23, 2016.
81. Sweden, Denmark, and Norway stopped funding the forest sector in Mozambique due to corruption issues. The FIP ended that situation, and these countries resumed their funding.
82. Fieldnotes, April 18, 2016.

Conclusion

1. Interview, March 18, 2016.

References

Abdenur, Adriana Erthal, and Danilo Marcondes de Souza Neto. 2013. "Brazil's Development Cooperation with Africa: What Role for Democracy and Human Rights?" *SUR—International Journal on Human Rights* 19: 17–35.

Agamben, Giorgio. 2009. "What Is an Apparatus?" In *What Is an Apparatus? And Other Essays*, 1–24. Stanford, CA: Stanford University Press.

Agrawal, Arun. 2005. *Environmentality: Technologies of Government and the Making of Subjects*. Durham, NC: Duke University Press.

Aksoy, M. Ataman, and Fahrettin Yagci. 2012. "Mozambique Cashew Reforms Revisited—Policy Research Working Paper 5939." World Bank—Poverty Reduction and Economic Management Network International Trade Department. https://ssrn.com/abstract=1982448.

Allitt, Patrick. 2014. *A Climate of Crisis: America in the Age of Environmentalism*. New York: Penguin Books.

Almeida, Mauro Barbosa. 2002. "The Politics of Amazonian Conservation: The Struggle of Rubber Tappers." *Journal of Latin American Anthropology* 7 (1): 170–219.

Almeida, Mauro Barbosa, Cristina Scheibe Wolff, Eliza Lozano Costa, and Mariana C. Pantoja Franco. 2002. "Habitantes: Os Seringueiros." In *Enciclopédia da Floresta: O Alto Juruá: Práticas e Conhecimento das Populações*, edited by Manuela Carneiro da Cunha and Mauro Barbosa de Almeida, 105–46. São Paulo: Companhia das Letras.

Amorim, Celso. 2013. *Breves Narrativas Diplomáticas*. São Paulo, Brazil: Benvirá.

Appadurai, Arjun. 1986. "Introduction: Commodities and the Politics of Value." In *The Social Life of Things: Commodities in Cultural Perspective*, edited by Arjun Appadurai, 3–63. Cambridge: Cambridge University Press.

Asiyanbi, Adeniyi P. 2016. "A Political Ecology of REDD+: Property Rights, Militarised Protectionism, and Carbonised Exclusion in Cross River." *Geoforum* 77: 146–56.

Asiyanbi, Adeniyi P., and Jens Friis Lund. 2020. "Policy Persistence: REDD+ between Stabilization and Contestation." *Journal of Political Ecology* 27: 378–495.

Austin, John. 1962. *How to Do Things with Words*. Cambridge, MA: Harvard University Press.

Baccini, A., S. J. Goetz, W. S. Walker, N. T. Laporte, M. Sun, D. Sulla-Menashe, J. Hackler, et al. 2012. "Estimated Carbon Dioxide Emissions from Tropical Deforestation Improved by Carbon-Density Maps." *Nature Climate Change*, no. 1354 (January): 1–4.

Bachram, Heidi. 2004. "Climate Fraud and Carbon Colonialism: The New Trade in Greenhouse Gases." *Capitalism Nature Socialism* 15 (4): 1–16.

Barnett, Michael, and Martha Finnemore. 2004. *Rules for the World: International Organizations in Global Politics*. Ithaca, NY: Cornell University Press.

Baron, Richard, and Michel Colombier. 2005. "Emissions Trading under the Kyoto Protocol: How Far from the Ideal?" In *Climate Change and Carbon Markets: A Handbook of Emission Reduction Mechanisms*, edited by Farhana Yamin, 153–65. Oxfordshire, UK: Earthscan from Routledge.

Baumert, Sophia, Ana Catarina Luz, Janet Fisher, Frank Vollmer, Casey M. Ryan, Genevieve Patenaude, Pedro Zorrilla-Miras, Luis Artur, Isilda Nhantumbo, and Duncan Macqueen. 2016. "Charcoal Supply Chains from Mabalane to Maputo: Who Benefits?" *Energy for Sustainable Development* 33: 129–38.

Beck, Ulrich. 1992. *Risk Society: Towards a New Modernity*. London: Sage Publication.

Berkman, Steve. 2008. *The World Bank and the Gods of Lending*. Boulder, CO: Kumarian Press.

Beymer-Farris, Betsy, and Thomas Bassett. 2012. "The REDD Menace: Resurgent Protectionism in Tanzania's Mangrove Forests." *Global Environmental Change* 22: 332–41.

Bodansky, Daniel. 1993. "The United Nations Framework Convention on Climate Change: A Commentary." *Yale Journal of International Law* 18 (2): 451–558.

Boon, James A. 1982. *Other Tribes, Other Scribes: Symbolic Anthropology in the Comparative Study of Cultures, Histories, Religions, and Texts*. Cambridge: Cambridge University Press.

Bourdieu, Pierre. 1977. *Outline of a Theory of Practice*. Cambridge: Cambridge University Press.

Bourdieu, Pierre. 1993. "The Field of Cultural Production, or: The Economic World Reversed." In *The Field of Cultural Production: Essays on Art and Literature*, edited by Randal Johnson. New York: Columbia University Press.

Bulkan, Janette. 2016. "The Limitations of International Auditing: The Case of the Norway-Guyana REDD+ Agreement." In *The Carbon Fix: Forest Carbon, Social Justice and Environmental Governance*, edited by Stephanie Paladino and Shirley Fiske, 91–106. London: Routledge.

Brenneis, Don. 2006. "Reforming Promise." In *Documents—Artifacts of Modern Knowledge*, edited by Annelise Riles. Ann Arbor: The University of Michigan Press.

Brenner, Neil, Jamie Peck, and Nik Theodore. 2010. "Variegated Neoliberalization: Geographies, Modalities, Pathways." *Global Networks* 10 (2): 182–222.

Briggs, Charles L. 1998. "'You're a Liar—You're Just Like a Woman!' Constructing Dominant Ideologies of Language in Warao Men's Gossip." In *Language Ideologies: Practice and Theory*, edited by Bambi B. Shieffelin, Kathryn A. Woolard, and Paul V. Kroskrity. Cambridge: Cambridge University Press.

Brockhaus, Maria, Grace Wong, Cecilia Luttrell, Lasse Loft, Thuy Thu Pham, Amy E. Duchelle, Samuel Assembe-Mvondo, and Monica Di Gregorio. 2014. "Operationalizing Safeguards in National REDD+ Benefit-Sharing Systems." Center for International Forestry Research (CIFOR). http://www.cifor.org/publications/pdf_files/SafeguardBrief/5187-brief.pdf.

Bumpus, Adam G. 2011. "The Matter of Carbon: Understanding the Materiality of TCO2e in Carbon Offsets." *Antipode* 43 (3): 612–38.

Bumpus, Adam G., and Diana M. Liverman. 2008. "Accumulation by Decarbonization and the Governance of Carbon Offsets." *Economic Geography* 84 (2): 127–55.

Büscher, Bram. 2013. "Selling Success: Constructing Value in Conservation and Development." *World Development* 57: 79–90.

Büscher, Bram, and Robert Fletcher. 2018. "Under Pressure: Conceptualising Political Ecologies of Green Wars." *Conservation & Society* 16 (2): 105–13.

Cabello, Joanna, and Tamra Gilbertson. 2011. "No REDD! A Reader: A Collection of Articles Written by REDD Monitor, Global Justice Ecology Project, Censat Agua Viva, Amazon Watch, Acción Ecológica, COECOCEIBA-AT, OFRANEH, World Rainforest Movement, Carbon Trade Watch, Rising Tide, ETC Group and Indigenous Environmental Network." Carbon Trade Watch and Indigenous Environmental Network. http://no-redd.com/wp-content/uploads/2015/01/REDDreadererEN.pdf.

Cabral, Lídia, and Alex Shankland. 2013. "Narratives of Brazil-Africa Cooperation for Agricultural Development: New Paradigms?" China and Brazil in Africa Agriculture (CBAA) Project Working Paper 51.

Callison, Candis. 2014. *How Climate Change Comes to Matter: The Communal Life of Facts*. Durham, NC: Duke University Press.

Callon, Michel. 1998. *The Laws of the Markets*. Oxford, UK: Blackwell Publishing.

Campbell, Jeremy M. 2015. *Conjuring Property: Speculation & Environmental Futures in the Brazilian Amazon*. Seattle: University of Washington Press.

Carbon Pricing Leadership Coalition (CPLC). 2017. "Report of the High-Level Commission Carbon Prices." World Bank. https://static1.squarespace.com/static/54ff9c5ce4b0a53decccfb4c/t/59244eed17bffc0ac256cf16/1495551740633/Carbon-Pricing_Final_May29.pdf.

Cardoso, Júlio Gardé Alfaro. 1946. *O Problema Florestal de Moçambique*. Lourenço Marques: Sociedade de Estudos da Colónia de Moçambique.

Carr, Mathew. 2013. "UN Emission Credits Surge as Developers Delay Carbon Claims." *Bloomberg.* http://www.bloomberg.com/news/2013-04-09/un-emission-credits-surge-as-developers-delay-carbon-cut-claims.html.

Carrier, James G., and Daniel Miller. 1998. *Virtualism: A New Political Economy.* Oxford: Berg Publishers.

Carrington, Damian. 2010. "WikiLeaks Cables Reveal How US Manipulated Climate Accord." *Guardian*, December 3, 2010. http://www.guardian.co.uk/environment/2010/dec/03/wikileaks-us-manipulated-climate-accord.

Castree, Noel. 2003. "Commodifying What Nature?" *Progress in Human Geography* 27 (3): 273–97.

Castree, Noel. 2008. "Neoliberalising Nature: The Logics of Deregulation and Reregulation." *Environment and Planning A: Economy and Space* 40: 131–52.

Cesarino, Letícia. 2017. "Anthropology and the South-South Encounter: On 'Culture' in Brazil-Africa Relations." *American Anthropologist* 119 (2): 333–41.

Chazdon, Robin L., Pedro H. S. Brancalion, Lars Laestadius, Aoife Bennett-Curry, Kathleen Buckingham, Chetan Kumar, Julian Moll-Rocek, Ima Célia Guimarães Vieira, and Sarah Jane Wilson. 2016. "When Is a Forest a Forest? Forest Concepts and Definitions in the Era of Forest and Landscape Restoration." *Ambio* 45: 538–50.

Cherrett, I., J. Kirkby, O. Marleyn, B. Munslow, and P. O'Keefe. 1990. "Norwegian Aid and the Environment in Mozambique—An Identification of the Issues Mozambique Country Study and Norwegian Aid Review Special Study." ETC (UK). Biblioteca do Centro de Estudos Africanos, Universidade Eduardo Mondlane, Cota 169E.

Cherry, Todd L., Jon Hovi, and David M. McEvoy. 2014. "Introduction." In *Toward a New Climate Agreement: Conflict, Resolution and Governance*, edited by Todd L. Cherry, Jon Hovi, and David M. McEvoy, ixx–xxix. Routledge Advances in Climate Change Research. London: Routledge.

Chitará, Sérgio. 2003. "Instrumentos Para a Promoção Do Investimento Privado Na Indústria Florestal Moçambicana." Ministério da Agricultura e Desenvolvimento Rural-Direcção Nacional de Florestas e Fauna Bravia. https://pubs.iied.org/pdfs/13517IIED.pdf.

Chomba, Susan, Juliet Kariuki, Jens Friis Lund, and Fergus Sinclair. 2016. "Roots of Inequity: How the Implementation of REDD+ Reinforces Past Injustices." *Land Use Policy* 50: 202–13.

Cicalo, André. 2014. "From Racial Mixture to Black Nation: Racialising Discourses in Brazil's African Affairs." *Journal of the Society for Latin American Studies* 33 (1): 16–30.

Clarke, John, Dave Bainton, Noémi Lendvai, and Paul Stubbs. 2015. *Making Policy Move: Towards a Politics of Translation and Assemblage.* Bristol, UK: Policy Press.

Climate Focus. 2013. "Acre, Brazil: Subnational Leader in REDD+." Climate Focus. http://www.climatefocus.com/sites/default/files/acre_brazil.pdf.

Climate Summit. 2014. "Report on the UN Climate Summit Forests Action Area Plenary Session." Climate Summit 2014: Catalyzing Action. http://www.uncclearn.org/sites/default/files/inventory/undp202.pdf.

Colchester, Marcus, and Larry Lohmann. 1990. *The Tropical Forestry Action Plan: What Progress?* Penang, Malaysia: World Rainforest Movement and the Ecologist.

Coles, Kimberley. 2007. *Democratic Designs: International Intervention and Electoral Practices in Postwar Bosnia-Herzegovina*. Ann Arbor: The University of Michigan Press.

Comaroff, John, and Jean Comaroff, eds. 1999. *Civil Society and the Political Imagination in Africa—Critical Perspectives*. Chicago: The University of Chicago Press.

Comissão Técnica de Planeamento e Integração Económica da Província de Moçambique. 1966. *III Plano de Fomento, Parte II, Relatórios Sectoriais, Volume I Agricultura e Silvicultura, Tomo VI Capítulos 18º e 19º*. Lisbon, Portugal: Ministério do Ultramar.

Cordeiro, Ana Dias. 2013 "'Moçambique Vive Uma Situação de Guerra Não Declarada." *PÚBLICO*. http://www.publico.pt/n1610722.

Coronil, Fernando. 1997. *The Magical State: Nature, Money, and Modernity in Venezuela*. Chicago: The University of Chicago Press.

Crewe, Emma, and Elizabeth Harrison. 1998. *Whose Development? An Ethnography of Aid*. London: Zed Books.

Cunha, Manuela Carneiro. 2012. *Índios No Brasil: História, Direitos e Cidadania*. São Paulo, Brazil: Editora Claro Enigma.

Cunha, Manuela Carneiro da. 2009. *"Culture" and Culture: Traditional Knowledge and Intellectual Rights*. Chicago: Prickly Paradigm Press.

Daviet, Florence, Gaia Larsen, Donna Lee, Stephanie Roe, Robert O'Sullivan, and Charlotte Streck. 2013. "Safeguards for REDD+ from a Donor Perspective." Climate Focus. http://www.fcmcglobal.org/documents/Climate_Focus_Safeguards_Report.pdf.

Dávila, Jerry. 2010. *Hotel Trópico: Brazil and the Challenge of African Decolonization, 1950–1980*. Durham, NC: Duke University Press.

Demeritt, David. 2001. "The Construction of Global Warming and the Politics of Science." *Annals of the Association of American Geographers* 91 (2): 307–37.

Dempsey, Jessica, and Daniel Chiu Suarez. 2016. "Arrested Development? The Promises and Paradoxes of 'Selling Nature to Save It.'" *Annals of the American Association of Geographers* 106 (3): 653–71.

Dent, Alexander S. 2013. "Intellectual Property in Practice: Filtering Testimony at the United States Trade Representative." *Journal of Linguistic Anthropology* 23 (2): E48–65.

Dent, Alexander S. 2016. "Policing the Unstable Materialities of Digital-Media Piracy in Brazil." *American Ethnologist* 43 (3): 424–36.

Depledge, Joanna. 2007. "A Special Relationship: Chairpersons and the Secretariat in the Climate Change Negotiations." *Global Environmental Politics* 7 (1): 45–68.

DeShazo, Jessica L., Chandra Lal Pandey, and Zachary A. Smith. 2016. *Why REDD Will Fail*. New York: Routledge.

Diallo, Rozenn Nakanabo. 2014. "Les Paradoxes Du Régime de l'Aide, Entre Injonctions Internationales et Logiques Nationales. Les Cas d'un Enclave Bureaucratique Au Mozambique." *Mondes En Développement* 42 (1/165): 51–63.

Disney, M. I., M. Boni Vicari, A. Burt, K. Calders, S. L. Lewis, P. Raumonen, and P. Wilkes. 2018. "Weighing Trees with Lasers: Advances, Challenges and Opportunities." *Interface Focus* 8 (2): 1–14.

Dooley, Kate. 2010. "Forest Watch Special Report–UNFCCC Climate Talks, 7–18 December 2009." FERN. http://www.fern.org/sites/fern.org/files/Copenhagen%20update.pdf.

Dooley, Kate. 2014. "Misleading Numbers: The Case for Separating Land and Fossil Based Carbon Emissions." FERN. http://www.fern.org/sites/fern.org/files/misleadingnumbers_full%20report.pdf.

Drigo, R., C. Cuambe, M. Lorenzini, A. Marzoli, J. Macuacua, C. Banze, P. Mugas, and D. Cunhete. 2008. "WISDOM Mozambique: Wood Energy Supply/Demand Analysis Applying the WISDOM Methodology." República de Moçambique, Ministério de Agricultura, Direcção Nacional de Terras e Florestas. http://www.wisdomprojects.net/global/csdetail.asp?id=16#.

Duchelle, Amy E., Marina Cromberg, Maria Fernanda Gebara, Raissa Guerra, Tadeu Melo, Anne M. Larson, Peter Cronkleton, et al. 2014. "Linking Forest Tenure Reform, Environmental Compliance, and Incentives: Lessons from REDD+ Initiatives in the Brazilian Amazon." *World Development* 55: 53–67.

Edwards, Paul N. 2001. "Representing the Global Atmosphere: Computer Models, Data, and Knowledge about Climate Change." In *Changing the Atmosphere: Expert Knowledge and Environmental Governance*, edited by Clark A. Miller and Paul N. Edwards, 31–65. Cambridge, MA: The MIT Press.

Edwards, Paul N., and Stephen H. Schneider. 2001. "Self-Governance and Peer Review in Science-for-Policy: The Case of the IPCC Second Assessment Report." In *Changing the Atmosphere: Expert Knowledge and Environmental Governance*, edited by Clark A. Miller and Paul N. Edwards, 219–46. Cambridge, MA: The MIT Press.

Edwards, Sophie. 2016. "New Acceleration Fund to Drive Investment in Forests." Devex. November 11, 2016. https://www.devex.com/news/sponsored/new-acceleration-fund-to-drive-investment-in-forests-89115.

Eisenhammer, Stephen. 2021. "The Amazon's Little Tipping Points: Nearing a Point of No Return?" Reuters. https://www.reuters.com/investigates/special-report/climate-un-amazon-tipping-point/.

Ekman, Sigrid-Marianella Stensrud, Huang Wenbin, and Ercilio Langa. 2013. *Chinese Trade and Investment in the Mozambican Timber Industry A Case Study from Cabo Delgado Province*. Bogor, Indonesia: Center for International Forestry Research (CIFOR).

Elsworth, Rob, Bryony Worthington, Michael Buick, and Patrick Craston. 2011. "Carbon Fat Cats 2011—The Companies Profiting from the EU Emissions Trading

Scheme." Sandbag. http://www.sandbag.org.uk/site_media/pdfs/reports/Sandbag_2011-06_fatcats.pdf.

Englund, Harri. 2006. *Prisoners of Freedom: Human Rights and the African Poor*. Berkeley: University of California Press.

Environmental Investigation Agency (EIA). 2013. "First Class Connections: Log Smuggling, Illegal Logging and Corruption in Mozambique." Environmental Investigation Agency (EIA). http://www.eia-international.org/wp-content/uploads/EIA-First-Class-Connections.pdf.

Escobar, Arturo. 1995. *Encountering Development: The Making and Unmaking of the Third World*. Princeton, NJ: Princeton University Press.

Fairbairn, Madeleine. 2013. "Indirect Dispossession: Domestic Power Imbalances and Foreign Access to Land in Mozambique." *Development and Change* 44 (2): 335–56.

Federative Republic of Brazil. 2015. "Intended Nationally Determined Contribution Towards Achieving the Objective of the United Nations Framework Convention on Climate Change." Federative Republic of Brazil. http://www4.unfccc.int/submissions/INDC/Published%20Documents/Brazil/1/BRAZIL%20iNDC%20english%20FINAL.pdf.

Feijó, João. 2021. "From the 'Faceless Enemy' to the Hypothesis of Dialogue: Identities, Pretensions and Channels of Communication with the *Machababs*." Observatório do Meio Rural. https://omrmz.org/omrweb/wp-content/uploads/DR-130-Cabo-Delgado-Pt-e-Eng.pdf.

Feldman, Ilana. 2008. *Governing Gaza—Bureaucracy, Authority, and the Work of Rule, 1917–1967*. Durham, NC: Duke University Press.

Ferguson, James. 1994. *The Anti-Politics Machine: "Development," Depoliticization, and Bureaucratic Power in Lesotho*. Minneapolis: University of Minnesota Press.

Ferguson, James. 2015. *Give a Man a Fish: Reflections on the New Politics of Distribution*. Durham, NC: Duke University Press.

Ferguson, James, and Akhil Gupta. 2002. "Spatializing States: Toward an Ethnography of Neoliberal Governmentality." *American Ethnologist* 29 (4): 981–1002.

FERN. 2012. "EU Forest Watch: Informing NGOs, MEPs, Member States, the European Commission and the Media." FERN. http://www.fern.org/sites/fern.org/files/FW%20177%20December%202012_1.pdf.

FERN. 2015. "EU Forest Watch Issue 210—December 2015: Forests in Spotlight as COP21 Climate Talks Approach Endgame." FERN. http://www.fern.org/node/5998.

FIAN International / Hands off the Land Alliance. 2012. "The Human Rights Impacts of Tree Plantations in Niassa Province, Mozambique." FIAN International. https://www.tni.org/files/download/niassa_report-hi.pdf.

Fishbein, Greg, and Donna Lee. 2015. "Early Lessons from Jurisdictional REDD+ and Low Emissions Development Programs." World Bank and The Nature Conservancy. https://www.forestcarbonpartnership.org/sites/fcp/files/2015/January/REDD%2B_LED_web_high_res.pdf.

Fleming, James Rodger. 1998. *Historical Perspectives on Climate Change*. Oxford, UK: Oxford University Press.

Fletcher, Robert. 2010. "Neoliberal Environmentality: Towards a Poststructuralist Political Ecology of the Conservation Debate." *Conservation & Society* 8 (3): 171–81.

Fletcher, Robert, and Bram Büscher. 2017. "The PES Conceit: Revisiting the Relationship between Payments for Environmental Services and Neoliberal Conservation." *Ecological Economics* 132: 224–31.

Fletcher, Robert, Wolfram Dressler, Bram Buscher, and Zachary R. Anderson. 2016. "Questioning REDD+ and the Future of Market-Based Conservation." *Conservation Biology* 30 (3): 673–75.

Flew, Terry. 2014. "Six Theories of Neoliberalism." *Thesis Eleven* 122 (1): 49–71.

Florêncio, Fernando. 2005. *Ao Encontro Dos Mambos: Autoridades Tradicionais VaNdau e Estado Em Moçambique*. Lisbon, Portugal: Instituto de Ciências Sociais.

Fogel, Cathleen. 2005. "Biotic Carbon Sequestration and the Kyoto Protocol: The Construction of Global Knowledge by the Intergovernmental Panel on Climate Change." *International Environmental Agreements* 5: 191–210.

Forest Trends' Ecosystem Marketplace. 2013. "FCP: REDD+ Finance: Who's Counting?" July 25, 2013. http://www.forestcarbonportal.com/content/redd-finance-whos-counting.

Foucault, Michel. 1991. "Governmentality." In *The Foucault Effect: Studies in Governmentality*, 87–104. Chicago: The University of Chicago Press.

Freyre, Gilberto. 1933. *Casa grande & senzala*. Lisbon, Portugal: Livros do Brasil.

Friends of the Earth International. 2010. "REDD: The Realities in Black and White." Friends of the Earth International (FoEI). http://www.foei.org/wp-content/uploads/2014/01/REDD-ingles-final-17–11.pdf.

Gal, Susan. 1988. "The Political Economy of Code Choice." In *Codeswitching: Anthropological and Sociolinguistic Perspectives*, edited by Monica Heller, 245–64. Contributions to the Sociology of Language. Berlin, Germany: Mouton de Gruyter.

Gal, Susan. 2005. "Language Ideologies Compared: Metaphors of Public/Private." *Journal of Linguistic Anthropology* 15 (1): 23–37.

Ganti, Tejaswini. 2014. "Neoliberalism." *Annual Review of Anthropology* 43: 89–104.

Garfield, Seth. 2004. "A Nationalist Environment: Indians, Nature, and the Construction of the Xingu National Park in Brazil." *Luso-Brazilian Review* 41 (1): 139–67.

Garfield, Seth. 2013. *In Search of the Amazon: Brazil, the United States, and the Nature of a Region*. Durham, NC: Duke University Press.

Geertz, Clifford. 1973. *The Interpretation of Cultures*. New York: Basic Books.

Geffray, Christian. 1991. *A Causa das Armas: Antropologia Da Guerra Contemporânea Em Moçambique*. Porto, Portugal: Edições Afrontamento.

Gilbertson, Tamra, and Oscar Reyes. 2009. "Carbon Trading: How It Works and Why It Fails." *Critical Currents* 7. Uppsala, Sweden: Dag Hammarskjöld Foundation.

Global Witness. 2012. "Safeguarding REDD+ Finance: Ensuring Transparent and Accountable International Financial Flows." Global Witness. http://www.globalwitness.org/sites/default/files/library/Safeguarding%20REDD+%20Finance.pdf.

Goldenberg, Suzanne, and John Vidal. 2010a. "Cancún Climate Change Summit: Exorcising the Ghosts of Copenhagen." *The Guardian*, December 5, 2010, sec. Environment. http://www.theguardian.com/environment/2010/dec/05/cancun-climate-talks-ghosts-copenhagen.

Goldenberg, Suzanne, and John Vidal. 2010b. "US Envoy Rejects Suggestion That America Bribed Countries to Sign up to the Copenhagen Accord." *The Guardian*, December 6, 2010. http://www.guardian.co.uk/environment/2010/dec/06/wikileaks-todd-stern-copenhagen-accord.

Goldman, Michael. 2001. "Constructing an Environmental State: Eco-Governmentality and Other Transnational Practices of a 'Green' World Bank." *Social Problems* 48 (4): 499–523.

Goldstein, Allie, and Gloria Gonzalez. 2014. "Turning over a New Leaf State of the Forest Carbon Markets 2014." Ecosystem Marketplace. http://www.forest-trends.org/documents/files/doc_4770.pdf.

Golub, Alexander, Sabine Fuss, Ruben Lubowski, Jake Hiller, Nikolay Khabarov, Nicolas Koch, Andrey Krasovskii, et al. 2018. "Escaping the Climate Policy Uncertainty Trap: Options Contracts for REDD+." *Climate Policy* 18 (10): 1227–34.

Gomes e Sousa, A. 1955. "A Protecção da Natureza no Ultramar Português Especialmente em Moçambique." In *Vigésimo Quinto Aniversário 1955*, 3–23. Lourenço Marques: Sociedade de Estudos da Província de Moçambique.

Gore, Al. 2006. *An Inconvenient Truth: The Planetary Emergency of Global Warming and What We Can Do about It*. New York: Rodale Press.

Governo do Estado Acre. 2010. "Lei Nº 2.308 de 22 de Outubro de 2010 Cria o Sistema Estadual de Incentivos a Serviços Ambientais—SISA, o Programa de Incentivos Por Serviços Ambientais—ISA Carbono e Demais Programas de Serviços Ambientais e Produtos Ecossistêmicos Do Estado Do Acre e Dá Outras Providências." Governo do Estado do Acre.

Governo do Estado Acre. 2012. "Construção Participativa Da Lei Do Sistema de Incentivos a Serviços Ambientais—SISA Do Estado Do Acre." IMC Acre.

Graham, Peter, and Gustavo Silva-Chávez. 2016. "The Implications of the Paris Climate Agreement for Private Sector Roles in REDD+." Forest Trends. http://www.forest-trends.org/documents/files/doc_5305.pdf.

Greenfield, Patrick, Jonathan Watts, Phoebe Weston, and Fiona Harvey. 2021. "Cop26: World Leaders Agree Deal to End Deforestation." *The Guardian*, November 1, 2021, sec. Environment. https://www.theguardian.com/environment/2021/nov/01/biden-bolsonaro-and-xi-among-leaders-agreeing-to-end-deforestation-aoe.

Greenleaf, Maron. 2016. "2015 Rappaport Student Prize Winner Maron Greenleaf Interviewed by Amelia Moore." *Anthropology News*, June 2016.

Gusterson, Hugh. 1997. "Studying Up Revisited." *Political and Legal Anthropology Review* 20 (1): 114–19.

Guyer, Jane I. 1994. "The Spatial Dimensions of Civil Society in Africa: An Anthropologist Looks at Nigeria." In *Civil Society and the State in Africa*, edited by John

W. Harbeson, Donald Rothchild, Naomi Chazan. Boulder, CO: Lynne Rienner Publishers.

Hacking, Ian. 1999. *The Social Construction of What?* Cambridge, MA: Harvard University Press.

Halpin, Elizabeth A. 1990. "Indigenous Peoples and the Tropical Forestry Action Plan." World Resources Institute. http://pdf.usaid.gov/pdf_docs/PNACA790.pdf.

Hanks, William F. 2000. "Indexicality." *Journal of Linguistic Anthropology* 9 (1–2): 124–26.

Harvey, David. 2005. *A Brief History of Neoliberalism*. Oxford, UK: Oxford University Press.

Hecht, Susanna, and Alexander Cockburn. 1990. *The Fate of the Forest: Developers, Destroyers, and Defenders of the Amazon*. Chicago: The University of Chicago Press.

Hein, Jonas I. 2019. *The Political Ecology of REDD+ in Indonesia: Agrarian Conflicts and Forest Carbon*. Routledge Studies in Political Ecology. London: Routledge.

Heller, Monica. 1988. "Introduction." In *Codeswitching: Anthropological and Sociolinguistic Perspectives*, edited by Monica Heller, 1–24. Contributions to the Sociology of Language. Berlin, Germany: Mouton de Gruyter.

Herndon, Thomas, Michael Ash, and Robert Pollin. 2013. "Does High Public Debt Consistently Stifle Economic Growth? A Critique of Reinhart and Rogoff." *Political Economy Research Institute Working Paper Series* (322): 1–25.

Heyman, Josiah McC. 2004. "The Anthropology of Power-Wielding Bureaucracies." *Human Organization* 63 (4): 487–500.

Hilgartner, Stephen. 2000. *Science on Stage: Expert Advice as Public Drama*. Stanford, CA: Stanford University Press.

Hilhorst, Dorothea. 2003. *The Real World of NGOs: Discourses, Diversity and Development*. London: Zed Books.

Hill, Jane H. 1998. "'Today There Is No Respect': Nostalgia, 'Respect,' and Oppositional Discourse in Mexicano (Nahuatl) Language Ideology." In *Language Ideologies: Practice and Theory*, edited by Bambi B. Shieffelin, Kathryn A. Woolard, and Paul V. Kroskrity. Cambridge: Cambridge University Press.

Ho, Karen. 2009. *Liquidated: An Ethnography of Wall Street*. Durham, NC: Duke University Press.

Hoag, Colin. 2011. "Assembling Partial Perspectives: Thoughts on the Anthropology of Bureaucracy." *Political and Legal Anthropology Review* 34 (1): 81–94.

Hoefle, Scott William. 2013. "Beyond Carbon Colonialism: Frontier Peasant Livelihoods, Spatial Mobility and Deforestation in the Brazilian Amazon." *Critique of Anthropology* 33 (2): 193–213.

Hoelle, Jeffrey. 2015. *Rainforest Cowboys: The Rise of Ranching and Cattle Culture in Western Amazonia*. Austin: University of Texas Press.

Howson, Peter, and Sara Kindon. 2015. "Analysing Access to the Local REDD+ Benefits of Sungai Lamandau, Central Kalimantan, Indonesia." *Asia Pacific Viewpoint* 56 (1): 96–110.

Hughes, David McDermott. 1998. "Policy Paper 44: Mapping the Hinterland: Land Rights, Timber, and Territorial Politics in Mozambique." University of California Institute on Global Conflict and Cooperation, University of California, San Diego, CA, September 1, 1998, 1–25.

Hughes, David McDermott. 2006. *From Enslavement to Environmentalism: Politics on a Southern African Frontier*. Seattle: University of Washington Press.

Hull, Matthew S. 2012. *Government of Paper: The Materiality of Bureaucracy in Urban Pakistan*. Berkeley: University of California Press.

Human Rights Watch. 2013. "'What Is a House Without Food?': Mozambique's Coal Mining Boom and Resettlements." Human Rights Watch. https://www.hrw.org/sites/default/files/reports/mozambique0513_Upload_0.pdf.

Iglesias, Marcelo Piedrafita. 2010. *Os Kaxinawá de Felizardo: Correrias, Trabalho e Civilização No Alto Juruá*. Brazil: Paralelo 15.

International Crisis Group (ICG). 2021. "Stemming the Insurrection in Mozambique's Cabo Delgado." Crisis Group. June 11, 2021. https://www.crisisgroup.org/africa/southern-africa/mozambique/303-stemming-insurrection-mozambiques-cabo-delgado.

Irvine, Judith T., and Susan Gal. 2000. "Language Ideology and Linguistic Differentiation." In *Regimes of Language*, edited by Paul Kroskrity, 35–84. Santa Fe, NM: School of American Research Press.

Isaacman, Allen. 1972. *Mozambique: The Africanization of a European Institution—The Zambezi Prazos, 1750–1902*. Madison: The University of Wisconsin Press.

Isaacman, Allen. 1996. *Cotton Is the Mother of Poverty: Peasants, Work, and Rural Struggle in Colonial Mozambique, 1938–1961*. Portsmouth, NH: Heinemann.

Jagger, Pamela, Maria Brockhaus, Amy E. Duchelle, Maria Fernanda Gebara, Kathleen Lawlor, Ida Aju Pradnja Resosudarmo, and William D. Sunderlin. 2014. "The Evolution of REDD+ Social Safeguards in Brazil, Indonesia and Tanzania." Center for International Forestry Research (CIFOR). http://www.cifor.org/publications/pdf_files/SafeguardBrief/5185-brief.pdf.

Jakobson. 1960. "Closing Statement: Linguistics and Poetics." In *Style in Language*. Cambridge, MA: MIT Press.

Jakobson. 1995. *On Language*. Cambridge, MA: Harvard University Press.

Jasanoff, Sheila. 1990. *The Fifth Branch: Science Advisers as Policymakers*. Cambridge, MA: Harvard University Press.

Jasanoff, Sheila, ed. 2004. *States of Knowledge: The Co-Production of Science and Social Order*. London: Routledge.

Jasanoff, Sheila, and Brian Wynne. 1998. "Science and Decisionmaking." In *Human Choice and Climate Change: The Societal Framework*, edited by Steve Rayner and Elizabeth L. Malone, 1–87. Vol. 1 of Human Choice and Climate Change. Columbus, OH: Battelle Press.

Jindal, Rohit, John M. Kerr, and Sarah Carter. 2012. "Reducing Poverty Through Carbon Forestry? Impacts of the N'hambita Community Carbon Project in Mozambique." *World Development* 40 (10): 2123–35.

Kashwan, Prakash, and Robert Holahan. 2014. "Nested Governance for Effective REDD+: Institutional and Political Arguments." *International Journal of the Commons* 8 (2): 554–75.

Keck, Margaret E. 1995. "Social Equity and Environmental Politics in Brazil: Lessons from the Rubber Tappers of Acre." *Comparative Politics* 27 (4): 409–24.

Kelty, Christopher. 2005. "Geeks, Social Imaginaries, and Recursive Publics." *Cultural Anthropology* 20 (2): 185–214.

Kill, Jutta. 2013. "Carbon Discredit: Why the EU Should Steer Clear of Forest Carbon Offsets." FERN and Les Amis de la Terre. http://www.fern.org/sites/fern.org/files/Nhambita_internet.pdf.

Klein, Naomi. 2014. *This Changes Everything: Capitalism vs. the Climate*. New York: Simon & Schuster.

Klein, Naomi. 2016. "Foreign Affairs Symposium" at Johns Hopkins University, Baltimore, MD, February 23, 2016.

Knorr-Cetina, Karin. 1999. *Epistemic Cultures: How the Sciences Make Knowledge*. Cambridge, MA: Harvard University Press.

Kossoy, Alexandre, Grzegorz Peszko, Klaus Oppermann, Nicolai Prytz, Noémie Klein, Kornelis Blok, Long Lam, et al. 2015. "State and Trends of Carbon Pricing 2015." Washington, DC: World Bank. http://www-wds.worldbank.org/external/default/WDSContentServer/WDSP/IB/2015/09/21/090224b0830f0f31/2_0/Rendered/PDF/State0and0trends0of0carbon0pricing02015.pdf.

Krech, Shepard III. 1999. *The Ecological Indian: Myth and History*. New York: W.W. Norton & Company, Inc.

Kutney, Gerald. 2014. *Carbon Politics and the Failure of the Kyoto Protocol*. Routledge Explorations in Environmental Studies. London: Routledge.

Kwa, Chunglin. 2001. "The Rise and Fall of Weather Modification: Changes in American Attitudes toward Technology, Nature, and Society." In *Changing the Atmosphere: Expert Knowledge and Environmental Governance*, edited by Clark A. Miller and Paul N. Edwards, 135–65. Cambridge, MA: MIT Press.

Kyed, Helene Maria. 2007. "State Recognition of Traditional Authority: Authority, Citizenship and State-Formation in Rural Post-War Mozambique." PhD diss., Roskilde University Centre.

Lane, Richard. 2015. "Resources for the Future, Resources for Growth: The Making of the 1975 Growth Ban." In *The Politics of Carbon Markets*, edited by Benjamin Stephan and Richard Lane, 27–50. London: Routledge.

Lang, Chris. 2013a. "The Warsaw Framework for REDD Plus: The Decision on National Forest Monitoring Systems." REDD-Monitor. http://www.redd-monitor.org/2013/12/10/the-warsaw-framework-for-redd-plus-the-decision-on-national-forest-monitoring-systems/.

Lang, Chris. 2013b. "The Warsaw Framework for REDD Plus: The Decision on REDD Finance." REDD-Monitor, November 29, 2013. http://www.redd-monitor.org/2013/11/29/the-warsaw-framework-for-redd-plus-the-decision-on-redd-finance/.

Lang, Chris. 2016. "REDD Is Dead. So Now, How Are We Going to Save the World's Forests?" *The Ecologist*. February 11, 2016. http://www.theecologist.org/blogs_and_comments/Blogs/2987097/redd_is_dead_so_now_how_are_we_going_to_save_the_worlds_forests.html.

Larson, Anne M., Maria Brockhaus, William D. Sunderlin, Amy E. Duchelle, Andrea Babon, Therese Dokken, Thuy Thu Pham, et al. 2013. "Land Tenure and REDD+: The Good, the Bad and the Ugly." *Global Environmental Change* 23: 678–89.

Latour, Bruno. 1987. *Science in Action: How to Follow Scientists and Engineers through Society*. Cambridge, MA: Harvard University Press.

Leach, Melissa, and Ian Scoones. 2015. *Carbon Conflicts and Forest Landscapes in Africa*. London: Routledge.

Lee, Benjamin, and Edward LiPuma. 2002. "Cultures of Circulation: The Imaginations of Modernity." *Public Culture* 14 (1): 191–213.

Lévi-Strauss, Claude. 1955. "The Structural Study of Myth." *The Journal of American Folklore* 68 (270): 428–44.

Lévi-Strauss, Claude. 1964. *The Raw and the Cooked: Introduction to a Science of Mythology*. Harmondsworth, England: Penguin Books.

Li, Tania. 2005. "Beyond 'The State' and Failed Schemes." *American Anthropologist* 107 (3): 383–94.

Li, Tania. 2007. "Practices of Assemblage and Community Forest Management." *Economy and Society* 36 (2): 263–93.

LiPuma, Edward. 2017. *The Social Life of Financial Derivatives: Markets, Risk, and Time*. Durham, NC: Duke University Press.

LiPuma, Edward, and Benjamin Lee. 2012. "A Social Approach to the Financial Derivatives Markets." *South Atlantic Quarterly* 111 (2): 289–316.

Lohmann, Larry. 2008. "Carbon Trading, Climate Justice and the Production of Ignorance: Ten Examples." *Development* 51 (3): 359–65.

Lohmann, Larry. 2009. "Toward a Different Debate in Environmental Accounting: The Cases of Carbon and Cost-Benefit." *Organizations and Society* 34: 499–534.

Long, Andrew. 2013. "REDD+, Adaptation, and Sustainable Forest Management: Toward Effective Polycentric Global Forest Governance." *Tropical Conservation Science* 6 (3): 384–408.

Long, Stephanie, Ellen Roberts, and Julia Dehm. 2010. "Climate Justice Inside and Outside the UNFCCC: The Example of REDD." *The Journal of Australian Political Economy* 66: 222–46.

Lovell, Heather, Harriet Bulkeley, and Diana Liverman. 2009. "Carbon Offsetting: Sustaining Consumption?" *Environment and Planning A: Economy and Space* 41: 2357–79.

Lovell, Heather, and Donald MacKenzie. 2015. "Allometric Equations and Timber Markets: An Important Forerunner of REDD+?" In *The Politics of Carbon Markets*, edited by Benjamin Stephan and Richard Lane, 69–90. London: Routledge.

Lovera, Simone. 2013. "Guest Post: A Pathetic REDD Package." REDD-Monitor, December 3, 2013. http://www.redd-monitor.org/2013/12/03/guest-post-a-pathetic-redd-package/.

Lubkemann, Stephen C. 2008. *Culture in Chaos: An Anthropology of the Social Condition in War*. Chicago: The University of Chicago Press.

Lund, Jens Friis, Eliezeri Sungusia, Mathew Bukhi Mabele, and Andreas Scheba. 2017. "Promising Change, Delivering Continuity: REDD+ as Conservation Fad." *World Development* 89: 124–39.

Lunstrum, Elizabeth. 2008. "Mozambique, Neoliberal Land Reform, and the Limpopo National Park." *Geographical Review* 98: 339–55.

Lunstrum, Elizabeth. 2010. "Reconstructing History, Grounding Claims to Space: History, Memory, and Displacement in the Great Limpopo Transfrontier Park." *South African Geographical Journal* 92 (2): 129–43.

Machado, Altino. 2008. "O Extrativismo Está Falido." *Blog do Altino Machado* (blog). December 19, 2008. http://www.altinomachado.com.br/search?updated-max=2008-12-20T04:32:00-08:00&max-results=20.

Machaqueiro, Raquel. 2017. "The Semiotics of Carbon: Atmospheric Space, Fungibility, and the Production of Scarcity." *Economic Anthropology* 4: 82–93.

Machaqueiro, Raquel Rodrigues. 2019. "Environmentality by the United Nations Framework Convention for Climate Change: Neoliberal Ethos and the Production of Environmental Subjects in Acre and Mozambique." *Environment and Planning E: Nature and Space* 3 (2): 442–61.

MacKenzie, Catherine. 2006. "Forest Governance in Zambézia, Mozambique: Chinese Takeaway! Final Report for FONGZA." Tanzania Forest Conservation Group. http://coastalforests.tfcg.org/pubs/GovernanceZambezia-MZQ.pdf.

MacKenzie, Donald. 2009a. *Material Markets: How Economic Agents Are Constructed*. Oxford, UK: Oxford University Press.

MacKenzie, Donald. 2009b. "Making Things the Same: Gases, Emission Rights and the Politics of Carbon Markets." *Organizations and Society* 34: 440–55.

Mahanty, Sango, Sarah Milne, Wolfram Dressler, and Colin Filer. 2012. "The Social Life of Forest Carbon: Property and Politics in the Production of a New Commodity." *Human Ecology* 40 (5): 661–64.

Maniates, Michael F. 2001. "Individualization: Plant a Tree, Buy a Bike, Save the World?" *Global Environmental Politics* 1 (3): 31–52.

Marcu, Andrei. 2016. "Carbon Market Provisions in the Paris Agreement (Article 6)." Centre for European Policy Studies Special Report 128, Brussels, Belgium. https://www.ceps.eu/system/files/SR%20No%20128%20ACM%20Post%20COP21%20Analysis%20of%20Article%206.pdf.

Marcus, George E. 1995. "Ethnography in/of the World System: The Emergence of Multi-Sited Ethnography." *Annual Review of Anthropology* 24: 95–117.

Marx, Karl. (1867) 1990. *Capital Volume I*. London: Penguin Books.

Masco, Joseph. 2010. "Bad Weather: On Planetary Crisis." *Social Studies of Science* 40 (1): 7–40.

McAfee, Kathleen. 2017. "Profits and Promises: Can Carbon Trading Save Forests and Aid Development." In *The Carbon Fix: Forest Carbon, Social Justice, and Envi-*

ronmental Governance, edited by Stephanie Paladino and Shirley J. Fiske, 37–59. New York: Routledge.

McNamara, Robert, and Filipe Ribeiro de Meneses. 2014. "A África Do Sul Face à 'descolonização Exemplar' Portuguesa." In *Portugal e o Fim Do Colonialismo. Dimensões Internacionais*, edited by Miguel Bandeira Jerónimo and António Costa Pinto, 135–54. Lisbon, Portugal: Edições 70.

McSweeney, Robert. 2014. "New Paper Raises Question of Tropical Forest Carbon Storage." Carbon Brief. December 15, 2014. https://www.carbonbrief.org/new-paper-raises-question-of-tropical-forest-carbon-storage.

Meinert, Lotte, and Susan Reynolds Whyte. 2014. "Epidemic Projectification: AIDS Responses in Uganda as Event and Process." *Cambridge Anthropology* 32 (1): 77–94.

Michaelowa, Axel. 2005. "Determination of Baselines and Additionality for the CDM: A Crucial Element of Credibility of the Climate Regime." In *Climate Change and Carbon Markets: A Handbook of Emission Reduction Mechanisms*, edited by Farhana Yamin, 289–304. Oxfordshire, UK: Earthscan from Routledge.

Miller, Clark A. 2004. "Climate Science and the Making of a Global Political Order." In *States of Knowledge: The Co-Production of Science and Social Order*, edited by Sheila Jasanoff. London: Routledge.

Miller, Clark A., and Paul N. Edwards. 2001. "Introduction: The Globalization of Climate Science and Climate Politics." In *Changing the Atmosphere: Expert Knowledge and Environmental Governance*, edited by Clark A. Miller and Paul N. Edwards, 1–30. Cambridge, MA: MIT Press.

Miller, Daniel. 2002. "Turning Callon the Right Way Up." *Economy and Society* 31 (2): 2018–2233.

Mintz, Sidney. 1985. *Sweetness and Power: The Place of Sugar in Modern History*. London: Penguin Books.

Mitchell, Timothy. 2011. *Carbon Democracy: Political Power in the Age of Oil*. London: Verso.

Moore, Donald S. 2005. *Suffering for Territory: Race, Place, and Power in Zimbabwe*. Durham, NC: Duke University Press.

Mosse, David. 2005. *Cultivating Development: An Ethnography of Aid Policy and Practice*. London: Pluto Press.

Mulungo, André. 2016. "Operadores Florestais Acusam Governo de Falta de Vontade Para Acabar Com Exploração Ilegal." *Canal de Moçambique*, March 23, 2016.

Myers-Scotton, Carol. 1993. *Social Motivations for Codeswitching: Evidence from Africa*. Oxford, UK: Clarendon Press.

Neto, Manoel José de Miranda. 1979. *O Dilema Da Amazônia*. Petrópolis, Brazil: Editora Vozes.

Netting, Robert McC. 1977. *Cultural Ecology*. Menlo Park, CA: Benjamin/Cummings Publishing Company.

Neumann, Iver B. 2012. *At Home with the Diplomats: Inside a European Foreign Ministry*. Ithaca, NY: Cornell University Press.

Neumann, Roderick P. 1998. *Imposing Wilderness: Struggles over Livelihood and Nature Preservation in Africa*. Berkeley: University of California Press.

Newitt, Malyn. 1995. *A History of Mozambique*. Bloomington: Indiana University Press.

Nhantumbo, Isilda. 2012. "South-South REDD: A Brazil-Mozambique Initiative for Zero Deforestation with Pan-African Relevance." IIED—International Institute for Environment and Development. http://pubs.iied.org/pdfs/G03585.pdf.

Nicolás, Elena Sánchez. 2021. "EU Commission 'failed' on Assessing Mercosur Trade Deal." EUobserver. March 22, 2021. https://euobserver.com/climate/151302.

Norway. 2012. "Norway's Submission to the UNFCCC on Views on Results-Based Finance for REDD+ (March 2012)." https://unfccc.int/files/bodies/awg-lca/application/pdf/norway_submission_on_results-based_finance_for_redd+_final.pdf.

Obarrio, Juan. 2014. *The Spirit of the Laws in Mozambique*. Chicago: The University of Chicago Press.

Oreskes, Naomi. 2004. "The Scientific Consensus on Climate Change." *Science* 306 (5702): 1686. https://doi.org/10.1126/science.1103618.

O'Riordan, Timothy, Chester L. Cooper, Andrew Jordan, Steve Rayner, Kenneth R. Richards, Paul Runci, and Shira Yoffe. 1998. "Institutional Frameworks for Political Action." In *Human Choice and Climate Change: The Societal Framework*, edited by Steve Rayner and Elizabeth L. Malone, 345–439. Vol. 1 of Human Choice and Climate Change. Columbus, OH: Battelle Press.

O'Sullivan, Arthur, and Steven Sheffrin. 2006. *Economics: Principles and Tools*. Upper Saddle River, NJ: Pearson Custom Publishing.

Paladino, Stephanie, and Shirley J. Fiske. 2017. *The Carbon Fix: Forest Carbon, Social Justice, and Environmental Governance*. New York and London: Routledge.

Pantoja, Mariana Ciavatta. 2014. "Kuntanawa: Ayahuasca, Ethnicity, and Culture." In *Ayahuasca Shamanism: In the Amazon and Beyond*, edited by Beatriz Caiuby Labate and Clancy Cavnar, 40–58. Oxford, UK: Oxford University Press.

Pantoja, Mariana Ciavatta, Eliza Lozano Costa, and Augusto Postigo. 2009. "A Presença Do Gado Em Reservas Extrativistas: Algumas Reflexões." *Revista Pós Ciências Sociais* 6 (12): 115–30.

Parker, Charlie, Andrew Mitchell, Mandar Trivedi, and Niki Mardas. 2009. "The Little REDD+ Book: An Updated Guide to Governmental and Non-Governmental Proposals for Reducing Emissions from Deforestation and Degradation." Global Canopy Programme. http://redd.unfccc.int/uploads/2_162_redd_20091201_gcp.pdf.

Paula, Elder Andrade de. 2016. *Seringueiros e Sindicatos: Um Povo de Floresta Em Busca de Liberdade*. Rio Branco, Acre: Nepan Editora.

Pearshouse, Richard. 2020. "Brazil: Accelerating Deforestation of Amazon a Direct Result of Bolsonaro's Policies." Amnesty International. December 2, 2020. https://www.amnesty.org/en/latest/press-release/2020/12/brazil-accelerating-deforestation-of-amazon-a-direct-result-of-bolsonaros-policies/.

Peck, Jamie, and Nik Theodore. 2015. *Fast Policy: Experimental Statecraft at the Thresholds of Neoliberalism*. Minneapolis: University of Minnesota Press.

Penvenne, Jeanne Marie. 2015. *Women, Migration & the Cashew Economy in Southern Mozambique: 1945–1975*. Suffolk, UK: Boydell & Brewer, James Currey.

Pfeil, Evy von. n.d. "REDD Early Movers (REM): Rewarding Pioneers in Forest Conservation; Financial Rewards for Successful Climate Change Mitigation." BMZ—Federal Ministry for Economic Cooperation and Development. Last accessed October 11, 2022. https://unfccc.int/files/cooperation_and_support/financial_mechanism/standing_committee/application/pdf/rem_wfc_09_15_final.pdf.

Phillips, Tom. 2019. "'War for Survival': Brazil's Amazon Tribes Despair as Land Raids Surge under Bolsonaro." The Guardian, October 2, 2019. https://www.theguardian.com/world/2019/oct/02/war-for-survival-brazils-amazon-tribes-despair-as-land-raids-surge-under-bolsonaro.

Polanyi, Karl. [1944] 2001. *The Great Transformation: The Political and Economic Origins of Our Time*. Boston: Beacon Press.

Povinelli, Elizabeth. 2002. *The Cunning of Recognition: Indigenous Alterities and the Making of Australian Multiculturalism*. Durham: Duke University Press.

Raffles, Hugh. 2002. *In Amazonia: A Natural History*. Princeton, NJ: Princeton University Press.

Reed, Stanley. 2013. "In European Union, Emissions Trade Is Sputtering." *New York Times*, February 20, 2013. http://www.nytimes.com/2013/02/21/business/energy-environment/21iht-green21.html.

Reinhart, Carmen, and Kenneth Rogoff. 2010. "Growth in a Time of Debt." *American Economic Review: Papers & Proceedings* 100 (2): 573–78.

República de Moçambique. 2013. "Regulamento dos Procedimentos para Aprovação de Projectos de Redução de Emissões por Desmatamento e Degradação Florestal (REDD+)." República de Moçambique.

República de Moçambique. 2016. "Estratégia Nacional para a Redução de Emissões de Desmatamento e Degradação Florestal, Conservação de Florestas e Aumento de Reservas de Carbono Através de Florestas (REDD+) 2016–2030." Ministério da Terra, Ambiente e Desenvolvimento Rural (MITADER).

República de Moçambique. n.d. "Estratégia de Redução de Emissões por Desmatamento e Degradação Florestal: Reduzir as Emissões de Carbono e a Pobreza Melhorando o Maneio das Florestas." Ministério para Coordenação da Acção Ambiental (MICOA).

República de Moçambique, and Forest Carbon Partnership Facility (FCPF). 2015. "Zambézia Integrated Landscapes Management Program: Emission Reductions Program Idea Note (ER-PIN) Mozambique."

Rich, Nathaniel. 2018. "Losing Earth: The Decade We Almost Stopped Climate Change." *New York Times*, August 1, 2018. https://www.nytimes.com/interactive/2018/08/01/magazine/climate-change-losing-earth.html.

Riles, Annelise. 2001. *The Network Inside Out*. Ann Arbor: The University of Michigan Press.

Riley, W. J., Q. Zhu, and J. Y. Tang. 2018. "Weaker Land-Climate Feedbacks from Nutrient Uptake during Photosynthesis-Inactive Periods." *Nature Climate Change* 8: 1002–6.

Rist, Gilbert. (1997) 2010. *The History of Development: From Western Origins to Global Faith*. London: Zed Books.

Roe, Emery. 1994. *Narrative Policy Analysis: Theory and Practice*. Durham, NC: Duke University Press.

SADCC. 1989. *Annual Energy Report*. London: SADCC.

Sahlins, Marshall. 1992. "The Economics of Develop-Man in the Pacific." *RES Anthropology and Aesthetics* 21: 12–25.

Sahlins, Marshall. 2005. *Culture in Practice: Selected Essays*. New York: Zone Books.

Sanders, Todd, and Harry G. West. 2003. "Power Revealed and Concealed in the New World Order." In *Transparency and Conspiracy: Ethnographies of Suspicion in the New World Order*, edited by Harry G. West and Todd Sanders. Durham, NC: Duke University Press.

Saraiva, José Flávio. 1993. "A Construção e Desconstrução Do Discurso Culturalista Na Política Africana Do Brasil." *Revista de Informação Legislativa* 30 (118): 219–36.

Scott, James. 1998. *Seeing Like a State: How Certain Schemes to Improve the Human Condition Have Failed*. New Haven, CT: Yale University Press.

Senda, Raul. 2016. "Abate Indiscriminado de Madeira Na Província Da Zambézia: Um Crime Antigo, Mas de Que Ninguém Se Dá Conta." *Savana*, March 25, 2016.

Serra, Carlos Manuel. 2013. "Transmissibilidade Dos Direitos de Uso e Aproveitamento Da Terra Em Moçambique." In *Dinâmicas Da Ocupação e Do Uso Da Terra Em Moçambique*, edited by Carlos Manuel Serra and João Carrilho, 51–74. Maputo, Mozambique: Escolar Editora.

Shah, Alpa. 2010. *In the Shadows of the State: Indigenous Politics, Environmentalism, and Insurgency in Jharkhand, India*. Durham, NC: Duke University Press.

Shankland, Alex, and Euclides Gonçalves. 2016. "Imagining Agricultural Development in South–South Cooperation: The Contestation and Transformation of ProSAVANA." *World Development* 81: 35–46.

Shankland, Alex, Euclides Gonçalves, and Arilson Favareto. 2016. "Social Movements, Agrarian Change and the Contestation of ProSAVANA in Mozambique and Brazil." Future Agricultures Consortium: China and Brazil in African Agriculture. CBAA Working Papers, November 2016. https://opendocs.ids.ac.uk/opendocs/bitstream/handle/123456789/12687/FAC_Working_Paper_137.pdf?sequence=1&isAllowed=y.

Shapin, Steven, and Simon Schaffer. 1985. *Leviathan and the Air-Pump: Hobbes, Boyle, and the Experimental Life*. Princeton, NJ: Princeton University Press.

Shore, Cris. 2011. "Espionage, Policy and the Art of Government: The British Secret Services and the War on Iraq." In *Policy Worlds: Anthropology and the Analysis of Contemporary Power*, edited by Cris Shore, Susan Wright, and Davide Però, 169–86. New York: Berghahn Books.

Shore, Cris, and Susan Wright, eds. 1997. "Policy: A New Field of Anthropology." In *Anthropology of Policy: Critical Perspectives on Governance and Power*. London: Routledge.

Shore, Cris, and Susan Wright. 2000. "Coercive Accountability: The Rise of Audit Culture in Higher Education." In *Audit Cultures: Anthropological Studies in Accountability, Ethics and the Academy*, edited by Marilyn Strathern. Oxford, UK: Routledge.

Shore, Cris, and Susan Wright. 2011. "Introduction: Conceptualising Policy: Technologies of Governance and the Politics of Visibility." In *Policy Worlds: Anthropology and the Analysis of Contemporary Power*, Cris Shore, Susan Wright, and Davide Però, 1–25. New York: Berghahn Books.

Shore, Cris, and Susan Wright. 2015. "Governing by Numbers: Audit Culture, Rankings and the New World Order." *Social Anthropology/Anthropologie Sociale* 23 (1): 22–28.

Silverstein, Michael. 1998. "The Uses and Utility of Ideology." In *Language Ideologies: Practice and Theory*, edited by Bambi B. Shieffelin, Kathryn A. Woolard, and Paul V. Kroskrity. Cambridge: Cambridge University Press.

Smith, David. 2012. "Boom Time for Mozambique, Once the Basket Case of Africa." *Guardian*, March 27, 2012. https://www.theguardian.com/world/2012/mar/27/mozambique-africa-energy-resources-bonanza.

Sodikoff, Genese. 2012. *Forest and Labor in Madagascar: From Colonial Concession to Global Biosphere*. Bloomington: Indiana University Press.

Sousa, Esteves de. 1950. *Considerações Acerca do Equilíbrio entre as Comunidades Florestais e o Ambiente em Moçambique*. Lisboa: Ministério das Colónias—Junta de Investigações Coloniais.

Strathern, Marilyn. 1996. "Cutting the Network." *The Journal of the Royal Anthropological Institute* 2 (3): 517–35.

Strathern, Marilyn, ed. 2000. "Introduction: New Accountabilities." In *Audit Cultures: Anthropological Studies in Accountability, Ethics and the Academy*. Oxford, UK: Routledge.

Svarstad, Hanne, and Tor A. Benjaminsen. 2017. "Nothing Succeeds Like Success Narratives: A Case of Conservation and Development in the Time of REDD." *Journal of Eastern African Studies* 11 (3): 482–505.

Temudo, Marina, and João M. N. Silva. 2011. "Agriculture and Forest Cover Changes in Post-War Mozambique." *Journal of Land Use Science*, 1–18.

Trexler, Mark. 2016. "Houston, We Have a Problem with Carbon Pricing and It's Not the One You Think!" *Ecosystem Marketplace* (blog). August 22, 2016. http://www.ecosystemmarketplace.com/articles/houston-problem-carbon-pricing-not-one-think/.

Turner, Terence. 2002. "Representation, Politics, and Cultural Imagination in Indigenous Video: General Points and Kayapo Examples." In *Media Worlds: Anthropology on New Terrain*, edited by Faye D. Ginsburg, Lila Abu-Lughod, and Brian Larkin. Berkeley: University of California Press.

Turnhout, Esther, Aarti Gupta, Janice Weatherley-Singh, Marjanneke J. Vijge, Jessica de Koning, Ingrid J. Visseren-Hamakers, Martin Herold, and Markus Lederer. 2017. "Envisioning REDD+ in a Post-Paris Era: Between Evolving Expectations and Current Practice." *WIREs Climate Change* 8: 1–13.

UNFCCC. 1992. "United Nations Framework Convention on Climate Change." United Nations. https://unfccc.int/resource/docs/convkp/conveng.pdf.

UNFCCC. 2011. "Report of the Conference of the Parties on Its Sixteenth Session, Held in Cancun from 29 November to 10 December 2010 Addendum Part Two: Action Taken by the Conference of the Parties at Its Sixteenth Session." UNFCCC. https://unfccc.int/resource/docs/2010/cop16/eng/07a01.pdf.

UNFCCC. 2013a. "Warsaw Outcomes." UNFCCC. https://unfccc.int/process/conferences/the-big-picture/milestones/outcomes-of-the-warsaw-conference.

UNFCCC. 2013b. "UN Climate Change Conference in Warsaw Keeps Governments on a Track towards 2015 Climate Agreement." UNFCCC.

UNFCCC. 2013c. "Press Release Governments in Warsaw Make Breakthrough in Agreements to Cut Greenhouse Gas Emissions from Deforestation." UNFCCC.

UNFCCC. 2014a. "Report of the Conference of the Parties on Its Nineteenth Session, Held in Warsaw from 11 to 23 November 2013." UNFCCC. http://unfccc.int/resource/docs/2013/cop19/eng/10a01.pdf.

UNFCCC. 2014b. "Views on Methodological Guidance for Non-Market-Based Approaches Related to the Implementation of the Activities Referred to in Decision 1/CP.16, Paragraph 70: Submissions from Parties and Admitted Observer Organizations." https://unfccc.int/resource/docs/2014/sbsta/eng/misc03.pdf.

UNFCCC. 2014c. "Views on the Issues Referred to in Decision 1/CP.18, Paragraph 40: Submissions from Parties and Admitted Observer Organizations." https://unfccc.int/resource/docs/2014/sbsta/eng/misc04.pdf.

UNFCCC. 2015. "Paris Agreement." UNFCCC. http://unfccc.int/files/essential_background/convention/application/pdf/english_paris_agreement.pdf.

UNFCCC. 2017. "UNFCCC EHandbook." UNFCCC. https://unfccc.int/resource/bigpicture/.

United States of America. 2015. "U.S. Submission on Elements of the 2015 Agreement." https://unfccc.int/files/documentation/submissions_from_parties/adp/application/pdf/u.s._submission_on_elements_of_the_2105_agreement.pdf.

University of Minnesota. 2016. "Secondary Tropical Forests Absorb Carbon at Higher Rate than Old-Growth Forests." ScienceDaily, February 8, 2016. https://www.sciencedaily.com/releases/2016/02/160208135436.htm.

van der Sleen, Peter, Peter Groenendijk, Mart Vlam, Niels P. R. Anten, Arnoud Boom, Frans Bongers, Thijs L. Pons, Gideon Terburg, and Pieter A. Zuidema. 2015. "No Growth Stimulation of Tropical Trees by 150 Years of CO2 Fertilization but Water-Use Efficiency Increased." *Nature Geoscience* 8: 24–28.

van der Veer, Peter. 2016. *The Value of Comparison*. Durham, NC: Duke University Press.

VCS. 2017. "Jurisdictional and Nested REDD+ (JNR) Requirements (VCS Version 3 Requirements Document)." VCS. http://database.v-c-s.org/sites/vcs.benfredaconsulting.com/files/Jurisdictional_and_Nested_REDD%2B_Requirements_v3.4.pdf.

Venicios, Marcos. 2019. "Após Demitir Funcionários, Peixes da Amazônia pede Recuperação Judicial para Evitar Falência." *ac24horas.com—Notícias do Acre* (blog). February 1, 2019. https://ac24horas.com/2019/02/01/apos-demitir-funcionarios-peixes-da-amazonia-pede-recuperacao-judicial-para-evitar-falencia/.

Vidal, John, and Suzanne Goldenberg. 2014. "Snowden Revelations of NSA Spying on Copenhagen Climate Talks Spark Anger." *Guardian*, January 30, 2014. http://www.theguardian.com/environment/2014/jan/30/snowden-nsa-spying-copenhagen-climate-talks.

Vidal, John, and Fiona Harvey. 2013. "Green Groups Walk out of UN Climate Talks." *The Guardian*, November 21, 2013, sec. Environment. http://www.theguardian.com/environment/2013/nov/21/mass-walk-out-un-climate-talks-warsaw.

Vieilledent, Ghislain, Fabian J. Fischer, Jérôme Chave, Daniel Guibal, Patrick Langbour, and Jean Gérard. 2018. "New Formula and Conversion Factor to Compute Basic Wood Density of Tree Species Using a Global Wood Technology Database." *American Journal of Botany* 105 (10): 1653–61.

von Hellermann, Pauline. 2013. *Things Fall Apart? The Political Ecology of Forest Governance in Southern Nigeria*. New York: Berghahn Books.

Walker, Michael Madison. 2012. "A Spatio-Temporal Mosaic of Land Use and Access in Central Mozambique." *Journal of Southern African Studies* 38 (3): 699–715.

Wallerstein, Immanuel. 2004. *World-Systems Analysis: An Introduction*. Durham, NC: Duke University Press.

Waugh, Linda R. 1980. "The Poetic Function in the Theory of Roman Jakobson." *Poetics Today* 2 (1): 57–82.

Weart, Spencer R. 2008. *The Discovery of Global Warming*. Cambridge, MA: Harvard University Press.

Weber, Max. (1946) 1973. *From Max Weber: Essays in Sociology*. Gerth and Mills edition. New York: Oxford University Press.

Wertz-Kanounnikoff, Sheila, and Desmond McNeill. 2012. "Performance Indicators and REDD+ Implementation." In *Analysing REDD+ Challenges and Choices*, edited by Arild Angelsen, Maria Brockhaus, William D. Sunderlin, and Louis V. Verchot, 233–46. Bogor, Indonesia: Center for International Forestry Research (CIFOR).

West, Harry G. 2008. "'Govern Yourselves!' Democracy and Carnage in Northern Mozambique." In *Democracy: Anthropological Approaches*, edited by Julia Paley. Santa Fe, NM: School for Advanced Research Press.

West, Paige. 2012. *From Modern Production to Imagined Primitive: The Social World of Coffee from Papua New Guinea*. Durham, NC: Duke University Press.

Woolard, Kathryn A. 1998. "Introduction: Language Ideology as a Field of Inquiry." In *Language Ideologies: Practice and Theory*, edited by Bambi B. Shieffelin, Kathryn A. Woolard, and Paul V. Kroskrity. Cambridge: Cambridge University Press.

World Bank. 1988. *Mozambique: Country Environmental Issues Paper*. Washington: World Bank.

World Bank, Ecofys, and Vivid Economics. 2016. "State and Trends on Carbon Pricing 2016." World Bank. http://documents.worldbank.org/curated/

en/598811476464765822/pdf/109157-REVISED-PUBLIC-wb-report-2016-complete-161214-cc2015-screen.pdf.

World Bank, and República de Moçambique. n.d. "Zambezia Integrated Landscape Management Program: Towards a Sustainable Forest Management and Improved Livelihoods of Rural Communities." World Bank IBRD-IDA.

World Rainforest Movement. 2016. "How Does the FAO Forest Definition Harm People and Forests? An Open Letter to the FAO." http://wrm.org.uy/other-relevant-information/how-does-the-fao-forest-definition-harm-people-and-forests-an-open-letter-to-the-fao/.

Wunder, Sven. 2005. "Payments for Environmental Services: Some Nuts and Bolts." CIFOR. http://www.cifor.org/publications/pdf_files/OccPapers/OP-42.pdf.

WWF-Brasil. 2013. "O Sistema de Incentivos Por Serviços Ambientais Do Estado Do Acre, Brasil: Lições Para Políticas, Programas e Estratégias de REDD Jurisdicional." WWF-Brasil. http://imc.ac.gov.br/wp-content/uploads/2016/09/O-SISA-Acre.pdf.

Yamin, Farhana, ed. 2005. *Climate Change and Carbon Markets: A Handbook of Emission Reduction Mechanisms*. Oxfordshire, UK: Earthscan from Routledge.

Yamin, Farhana, and Joanna Depledge. 2004. *The International Climate Change Regime: A Guide to Rules, Institutions and Procedures*. Cambridge: Cambridge University Press.

Young, Crawford. 1994. "In Search of Civil Society." In *Civil Society and the State in Africa*, edited by John Willis Harbeson, Donald S. Rothchild, and Naomi Chazan, 33–50. Boulder, CO: Lynne Rienner Publishers.

Zaloom, Caitlin. 2006. *Out of the Pits: Traders and Technology from Chicago to London*. Chicago: The University of Chicago Press.

Ziegler, Alan D., Jacob Phelps, Jia Qi Yuen, Edward L. Webb, Deborah Lawrence, Jeff M. Fox, Thilde B. Bruun, et al. 2012. "Carbon Outcomes of Major Land-Cover Transitions in SE Asia: Great Uncertainties and REDD+ Policy Implications." *Global Change Biology*, 1–13.

Zwick, Steve. 2016. "After Paris, Here's What Lies Ahead for Carbon Markets." *Ecosystem Marketplace* (blog). February 1, 2016. http://www.ecosystemmarketplace.com/articles/green-lights-and-speed-bumps-on-road-to-markets-under-paris-agreement/.

Zwick, Steve. 2017. "As REDD Talks Progress in Bonn, Question Arises: Who Can Use the Term 'REDD+'?" *Ecosystem Marketplace* (blog). November 14, 2017. http://www.ecosystemmarketplace.com/articles/as-redd-talks-progress-in-bonn-question-arises-who-can-use-the-term-redd/.

Zwick, Steve. 2018. "REDD Dawn: The 60-Year Evolution of Forest Carbon." *Ecosystem Marketplace* (blog). September 21, 2018. https://www.forest-trends.org/ecosystem_marketplace/redd-dawn.

Index

accuracy, 30, 88, 92, 93–96, 101, 109, 120, 123, 174, 248
additionality, 157, 208, 209, 215, 238, 293
afforestation, 59, 103, 105–7, 263
agriculture, conservation, 179, 182, 216, 217, 220–22, 225, 226, 232, 237, 238, 250, 278; shifting, 179, 180, 181, 225, 235; subsistence, 164, 170, 223, 256
agroforestry agents, 209, 211
Al Gore, 15, 35
allometric equation, 111, 278
Amazon, 10, 14, 24, 28, 90, 116, 145, 146, 173, 176, 191, 200, 203–6, 208, 209, 215, 253, 266–69; Amazon Fund, 24, 100, 115, 117, 118, 130, 152, 159, 175, 263, 264, 268, 272
"anti-politics machine," 17, 196, 243
apparatus, "environmentality," 161, 185, 186, 248, 249; governance, 184; institutional, 161; policy, 139, 155, 184, 185, 247

authority, 19, 40, 43, 64, 66, 246, 262; of certifiers, 124, 196; cognitive, 41; of the negotiation process within the UNFCCC, 80, 81, 84, 123; scientific, 12, 36, 41, 43, 45, 50, 63, 246; state, 166, 192, 212; transnational, 74, 248

benefits, 19, 37, 57, 93, 105, 130–32, 142, 169, 170, 199, 201, 212, 215, 217, 227, 229, 232, 240, 269; co-benefits, 129, 265; non-carbon, 129, 130, 135, 265
Bolsa Família. *See* Family Grant
Bolsa Floresta. *See* Forest Grant
Brundtland Report, 42, 68, 258, 260, 277
bureaucracy, 81; bureaucratic genre, 77, 80, 81, 92, 261; mode of knowledge production, 81, 84; power, 83; procedures, 196; rules, 123; style, 80

capillarity, 12, 24, 64, 189, 240, 243, 247; capillary, 242, 246

capitalism, 9, 138, 144, 209; green, 146, 147, 150
carbon accounting, 14, 73, 95, 107, 111, 120, 121, 123, 135, 157, 165, 167, 191, 192, 195, 197, 198, 213, 215, 250, 262; budget, 239; "colonialism," 49, 51; commoditization of, 30, 53, 54, 60, 162, 163, 183; credits, 107, 135, 160, 169, 212, 267, 271; cycle, 41; dioxide, 35, 42, 49, 50; emissions, 11, 35, 36, 37, 46, 47, 49, 61, 62, 63, 89, 95, 97, 156, 159, 161, 189, 195, 214, 217, 218, 226, 227, 229, 239, 240, 243, 250, 265; finance, 97, 269; footprint, 254; forest, 11, 19, 20, 61, 62, 72, 73, 97, 106, 109, 111–14, 120, 128, 139, 266, 270; fossil fuels, 186; internalization of, 245, 250; markets, 7, 10, 16, 24, 36, 40, 47, 49–51, 53–55, 57, 58, 60, 62, 63, 73, 98, 100, 108, 110, 115, 119, 131, 139, 140, 147, 163, 191, 236, 249, 250, 259, 266, 267; offsets, 11, 37, 50, 66, 119, 132, 139, 140, 153, 157, 159, 160, 162, 170, 172, 191, 192, 212–14, 217, 218, 237, 240, 265, 268, 269; ownership, 213; projects, 86, 97, 168; rights, 12, 164, 169, 170, 212, 213; sequestration, 170, 233, 235, 263; sinks, 10, 12, 66, 71, 105, 106, 123, 124, 126, 134, 264; social cost of, 56, 57, 60, 260; stocks, 11, 156, 158, 165, 178, 238; storage, 124, 199, 264; trading, 26, 49, 51, 54, 162, 163, 165, 168, 185, 237, 241; valuation, 10, 53, 55, 56, 58
cashew, 222, 223, 225, 250, 277
cattle, 190, 201, 203, 206–8, 253, 274; "culture," 207; herders, 207; pastures, 145; ranchers, 145, 199; ranching, 151, 198, 200, 203, 205, 207, 208, 215, 242, 252, 268, 274; stocking, 39, 207

CDSA (*Companhia de Desenvolvimento de Serviços Ambientais*, Company for the Development of Environmental Services), 153, 190, 268
Celso Correia, 3
CEVA (*Comissão Estadual de Validação e Acompanhamento*, State Commission of Validation and Monitoring), 153, 154, 194, 268
chain, commodity, 120, 205, 223, 225; "green supply," 222; of indebtedness, 204; production, 227; productive, 223; supply, 251; value, 223
charcoal, 179, 183, 226, 227, 277; consumption, 226; high-tech, 226; production, 164, 179–81, 220, 222, 226, 228, 239, 277, 278; sustainable, 226, 227, 239, 277
Chico Mendes, 28, 144–46, 151, 152, 199, 267
CIMI (*Conselho Indigenista Missionário*; Indigenist Missionary Council), 140, 266
civil society, 7, 68, 142, 148–50, 186, 267, 268
civil war (in Mozambique), 23, 163, 166, 230, 270, 277
Clean Development Mechanism (CDM), 48, 106, 107, 112, 258, 259, 263, 270
climate, accord, 58, 68, 99, 133, 160; change, 3, 4, 5–17, 21, 27–31, 35, 36, 40–47, 49, 51, 54–60, 62, 63–65, 67, 68, 72, 73, 76, 81, 83, 84, 89, 90, 91, 95, 97, 98, 100, 102, 104, 109, 112–14, 123, 126, 125, 133–35, 159, 182, 215, 240, 241, 243, 245–47, 251–54, 257, 258, 261, 262; community, 83, 95, 198; global, 37–39, 59, 62, 64, 65, 74, 77, 84, 102, 197, 246; governance, 5–8, 21, 29, 31, 65, 77, 102, 246–48; mitigation, 59, 109, 275; modeling, 38–41, 46, 52, 62;

Index

negotiations, 4, 6, 16, 68, 71, 72, 198, 252; policy, 41, 46, 49, 53, 59, 61, 74, 77, 104, 110, 111, 126, 132, 133, 253; policy field, 40, 45; science, 30, 36, 43, 45, 46, 49, 53, 54, 62, 63, 64, 246, 247

Climate Community and Biodiversity Alliance (CCBA), 140, 150, 191, 265, 267, 269

Coalition for Rainforest Nations (CfRN), 103, 107, 134, 135, 136, 263

code-switching, 83, 84, 261

commensurability, 49, 256

commoditization, of carbon, 30, 53, 54, 58, 60, 163; of nature, 12, 104, 147, 150, 211

community, climate, 83, 84, 95; epistemic, 40–42, 45, 49, 53, 63, 74, 79, 80, 101, 246; forest management, 147, 216, 229–32; international, 128; linguistic, 74, 261; scientific, 39, 42, 221

Conference of the Parties (COP), 65–68, 71–73, 79, 80, 82, 86, 93, 96, 98, 99, 101, 103, 106, 108–10, 116, 128, 133–35, 160, 248, 251, 260–62

confusion, 131, 240, 241, 242, 249; institutional, 189

consensus, 4, 77; building, 44, 45, 70, 72, 80, 261; consensual narrative, 75, 76, 81; scientific, 40, 43, 45, 261

consistency, 30, 92, 94–96, 101, 120, 123, 248

cookstoves, 179, 182, 226, 227, 277

deforestation, 11, 14–17, 24, 30, 37, 61, 62, 71, 73, 74, 85, 86, 88–93, 99, 100, 102–8, 112–16, 118, 125–27, 133, 134, 139, 140, 143, 145, 152, 157–59, 161, 164, 165, 169, 170, 176, 178–83, 185–87, 189, 191, 194, 196–99, 202, 205, 206, 208–11, 213, 214, 216–22, 225–28, 232–36, 238–40, 249–53, 263, 265, 266, 272, 275–77

depoliticization, depoliticizing effect, 8, 14, 17, 18, 92, 96, 123, 174, 182

differentiation, ideology of, 79; mechanism of, 84

DUAT (*Direito ao Uso e Aproveitamento da Terra*, Right to Use and Harness the Land), 171, 232, 234, 278

"ecosystem services," 129, 142, 156, 157, 160, 177, 178, 185, 208, 211, 266, 267

elite cabinet, 216, 242

emerging farmers, 223–25, 234

enrollment, 7, 17–19, 30, 45, 87, 101, 102, 150, 176, 237, 239, 243

EDF (Environmental Defense Fund), 60, 112, 113, 146, 150, 185, 260, 263, 267, 268

eucalyptus, 227, 232–39, 242, 277, 278

externality, 47, 56, 250

extractive, activities, 120, 183, 205; reserves, 144, 145, 147, 198, 199, 201, 202, 207, 214

Family Grant, 177, 272

FAS (*Fundação Amazônia Sustentável*; Amazonas Sustainable Foundation), 175–78, 185, 272

fish, farming, 157, 198, 205, 206, 214, 215, 242; factory, 198, 205, 206, 214, 252

flexible mechanisms, 7, 47, 48, 55, 58, 66, 73

Food and Agriculture Organization (FAO), 115, 125–27, 272

forest, carbon, 60–62, 66, 72, 73, 97, 100, 106, 109–15, 120, 124, 128, 131, 139, 186, 199, 264, 266, 268; of Amazonia, 10, 14, 28, 176; Code, 160, 187, 269, 270; communities, 29, 106, 152, 200, 230; conservation, 12, 19, 28, 92, 103, 105, 106, 108, 126, 128, 142, 162, 168, 172, 183,

187, 209, 235, 264; definition, 30, 106, 123, 125, 126, 243, 250, 262, 272, 276; degradation, 11, 89, 103, 126, 134, 179, 220, 277; forestry, 64, 112, 118, 122, 125, 131, 133, 232, 233, 258; governance, 23, 24, 138, 185; management, 24, 106, 118, 147, 152, 154, 158, 159, 179, 198, 199, 201, 202, 214, 222, 226, 230–32, 233, 237, 238, 242, 252, 267, 278; miombo, 4, 176; people, 164, 165, 167, 216; plantation, 48, 103, 157, 165, 232, 270; rainforest, 124, 176, 263; resources, 123, 124, 144, 164, 165, 168, 214, 230, 263; restoration; tropical, 15, 112, 176, 263, 264
Forest Carbon Partnership Facility (FCPF), 117, 118, 122, 173, 174, 178, 219, 232, 237, 272, 278
Forest Grant, 177, 272
Forest Investment Plan (FIP), 219, 232, 233, 278
Forests and the European Union Resource Network (FERN), 109, 134
Forest Trends-Ecosystem Marketplace, 60, 116, 130, 133, 136, 150, 185, 214, 266, 286
Frente de Libertação de Moçambique; Mozambique Liberation Front (FRELIMO), 166
Friends of the Earth, 109
Fundação Nacional do Índio, National Indian Foundation (FUNAI), 212, 275
fungibility, 50, 53, 258

geoengineering, 38, 257
German cooperation agency (GIZ), 152, 154, 268
Gilé National Park, 220, 221, 229, 264, 276
Global South, 8, 11, 13–15, 17, 18, 24, 26, 27, 36, 43, 49, 51, 64, 71–73, 89, 90–92, 100–2, 104, 115, 120, 126, 127, 136, 165, 182, 183, 197, 242, 243, 245, 246, 251, 252, 258
governance, environmental, 105; neoliberal, 4, 5, 36, 53; transnational, 8–10, 12–14, 18, 20, 21, 24–26, 28–31, 36, 64–66, 72, 102, 104, 127, 137, 139, 150, 161, 163, 167, 178, 182, 184, 186–88, 213, 215, 218, 236, 241–43, 246–51
"governmentality," 120, 122, 123, 130, 265
greenhouse gases (GHG), 9–11, 35, 36, 39, 41, 42, 45–47, 49, 50, 52, 53, 55–58, 69, 71, 103, 105, 131, 254, 257, 260

Huni Kuin, 211, 275

incentives, 129, 134, 140, 142, 143, 148, 150, 151, 179, 181, 185, 225, 259
indigenous, affairs, 210, 250; communities/people, 144, 152, 178, 198, 200, 208–13, 215, 230, 266, 274, 275; land, 72, 191, 198, 208, 209, 212, 220, 253; leaders, 140, 141, 211, 215, 275; movements, 210, 209; reserves, 145; rights, 211
Instituto de Mudanças Climáticas, Institute of Climate Change (IMC), 153, 154, 190, 193, 241
intangibility, 54, 162, 237
International Finance Corporation (IFC), 234, 278
IPCC (Intergovernmental Panel on Climate Change), 36, 43–46, 48–50, 52, 66, 86, 106, 257–59, 261
ISA Carbono, 154, 156–59
IUCN (International Union for Conservation of Nature), 142, 146, 150, 267, 268

JA! (*Justiça Ambiental*, Environmental Justice), 164, 165, 169, 170, 272

James Hansen, 15, 42
jurisdictional, program/REDD, 140, 143, 156, 173, 190–95, 197, 209, 212, 213, 215, 218, 219, 240, 266, 274

Kaxinawá, 275
Kevin Conrad, 107, 113, 135, 136
Kyoto Protocol, 5–9, 30, 47, 48, 55, 58, 59, 66–68, 71, 73, 93, 101, 103, 105–7, 110, 112, 128, 160, 258, 260, 261, 263

land, grabbing, 107, 165, 211, 256, 270; management, 59, 62; rights, 12, 164, 169, 170, 185, 191, 232, 273; sector, 59, 64, 119, 134, 136, 192, 229, 243; tenure, 12, 171, 191, 194, 195, 213; use, 105, 106, 126, 131, 133, 152, 170, 171, 179, 233, 234, 242, 271, 278
"landscape approach," 180, 189, 216, 218–23, 226, 229, 230–34, 236, 238–41, 251, 252, 276–78
language ideology, 73, 74, 78, 79, 81, 92, 96, 101, 248, 261
leakage, 83, 109, 140, 263
legitimacy, 7, 12, 13, 17–19, 30, 30, 40, 64, 65, 77, 102, 114, 141, 148–52, 156, 159, 169, 185, 186, 197, 215, 242, 243, 246, 248–50
Lula da Silva, 21–23, 272
Lusotropicalismo, 21, 22, 256

Marina Silva, 28
market, actors/agents, 48, 54–56, 58–60, 214; advocates, 104, 120, 132, 135, 250, 259; approaches, 8, 12, 23, 47, 49, 118, 121, 136, 155, 185, 186, 203, 218, 260; carbon, 10, 11, 16, 19, 24, 36, 40, 47–58, 60–63, 73, 98, 100, 108, 114–16, 119, 131, 139, 140, 147, 153, 159, 162, 163, 172, 185, 191, 197, 214, 215, 236, 249, 250, 259, 266; compliance, 48, 55, 61, 97, 258, 263, 271; conservation, 120, 273; failure, 47, 63; financial, 54, 61, 107, 150, 163, 258; global/international, 58, 66, 73, 90, 97, 114, 118–20, 122, 131, 139, 140, 147, 151, 153, 159, 162, 172, 197, 199, 206, 215, 253; ideology, 49, 54, 59; initiatives/instruments, 7, 23, 30, 49; logic, 9, 12, 53, 108, 130, 131, 139, 214; mechanisms, 5, 7, 49, 58, 73, 108, 109, 114, 134, 135, 155, 177, 209, 218, 236, 242, 245, 258; prices, 54–56, 60, 217, 222; project-based, 52, 53; rationale/rationality, 14, 51, 54; rules, 12, 104, 119, 120; sociality of, 59; as solution, 4, 16, 54, 55, 57, 73, 113, 133, 224, 263; voluntary, 48, 110, 115, 122, 130–32, 140, 192, 258, 271
messiness, institutional, 184, 249
Measuring, reporting, and verifying (MRV), 88, 89, 95, 96, 99–101, 109, 120–25, 127, 135, 262
Ministério da Agricultura, Ministry of Agriculture (MINAG), 163, 165, 167, 172, 174, 269
Ministério da Coordenação Ambiental, Ministry of Coordination of Environmental Affairs (MICOA), 163–65, 218, 270
misrecognition, 77, 176; co- 41, 63; politics of, 46, 53
monitoring systems, 74, 86, 87, 110, 114, 115, 128, 136, 168, 263, 274
myth, 14, 147, 148, 256, 266; counter- 146, 149; of origin, 141, 144, 146, 149, 186, 248

Nature Conservancy (TNC), The, 263
neoliberal, beliefs, 113, 120; ideology, 63, 105, 119; imperatives, 104, 123; logic, 121; mode of governance, 4, 5, 8, 36, 41, 53, 120; neoliberalism, 4, 96, 104, 105, 245; neoliberalization,

5, 12, 139, 236, 245; perspective, 5, 8, 133, 268; polymorphism, 177; principles, 5, 14, 36, 186
nesting, 79, 191–93, 195, 273
network, 14, 18, 25, 27, 29, 63, 99, 109, 149, 154, 155, 166, 167, 175, 177, 182, 184, 196, 225
N'hambita Project, 162, 167, 168, 172, 173, 231, 237, 269
no REDDs, 17, 144, 151, 167

offset, 11, 15, 19, 36, 37, 48, 50, 53, 58, 66, 73, 108, 110, 114, 115, 119, 122, 131, 132, 139, 140, 153, 157, 159–62, 170, 191–93, 196, 212–14, 217, 218, 237, 240, 258, 265, 268, 269, 273

Paris Accord, Agreement, 3–7, 56–60, 68, 101, 133–36, 160, 189, 192
permanence, of carbon sinks, 105, 106, 112, 263
policy, 5, 8, 9, 12–22, 24–28, 31, 35, 36, 41, 43, 44, 46, 49, 54, 56, 59–64, 95, 99–101, 104, 110, 111, 118–21, 130, 133–35, 140, 148, 152, 159, 163, 169, 183, 189, 198, 199, 202, 203, 207, 208, 210, 213, 215, 234, 245–47, 250, 253, 257, 258, 262, 264–66, 268; apparatus, 139, 184, 185; field, 40, 41, 45, 46, 53, 74, 77; policymakers, 9, 12, 14, 20, 30, 36, 39–45, 47, 51, 55, 57–60, 62, 63, 73, 155, 241, 252, 258, 262; policymaking, 8, 14, 16–19, 25, 27, 28, 30, 35, 45, 49, 59, 104, 122, 141, 155, 161, 163, 168, 169, 173, 184, 195, 212, 224, 242, 255, 275

quantification, 39, 132, 135, 210

readiness plan idea note (R-PIN), 83, 174, 272
readiness preparation proposal (R-PP), 122, 174, 272

reading obstacles, 30, 79, 95
recursive, 36, 41, 51, 62, 63, 71, 76, 77, 79, 80, 101, 120, 132, 246; recursion, 63, 77, 79, 81, 261; recursivity, 12, 30, 78, 79, 80, 92, 95, 101, 261
reference levels, 83, 85, 88, 93, 99, 101, 121, 136, 140, 192, 262
reforestation, 59, 103, 105–7, 198, 216, 232, 235, 263, 264
Resistência Nacional Moçambicana, Mozambican National Resistance (RENAMO), 166, 231
responsibility, 3–5, 7–9, 14, 16, 26, 38, 48, 69, 70, 90, 92, 100–2, 112–14, 133, 149, 165, 242, 252, 254, 258
rubber, tapper(s), 144–46, 199–208, 214, 274, 275
rural, development, 19, 23, 164, 179, 182, 216, 219–21, 225, 232, 239, 240; union, 145, 199, 200, 203

safeguards, 85, 87, 99, 120, 121, 128–32, 136, 140, 154, 169, 191, 193, 194, 265, 266, 274
Sistema de Incentivos a Serviços Ambientais, System of Incentives for Environmental Services (SISA), 140, 144, 147–58, 160, 161, 173, 190, 191, 194–97, 205, 207, 209–11, 213–15, 249, 268, 273, 275
slash-and-burn, 127, 143, 170, 181, 183, 220–22, 228, 239
South-South cooperation, 23–25, 175–78, 267, 272
sovereignty, 4, 89, 100, 107, 122, 192, 213, 253, 267
standardization, 29, 39, 46, 92, 93, 96, 100, 101, 120, 121, 123–30, 135, 156, 183, 195, 196, 248, 250, 256
Steve Schwartzman, 113, 267, 268
subjectification, effects, 84, 121, 122, 139, 155, 163, 183, 185, 236, 240, 245, 247, 249, 255

Index

sustainable, development, 19, 49, 144, 145, 147, 152, 153, 214, 258; forest management, 147, 152, 158, 198, 199, 201, 202, 214, 220, 222, 231, 233, 237, 242, 252, 278; sustainability, 26, 37, 142, 143, 157, 161, 168, 189, 205–7, 226, 227, 229, 239, 240
synecdoche, 30, 36, 42, 45, 53, 63, 245, 257; synecdochical re-inscription, 30, 35, 45, 49, 54, 60, 62–64

template, international template, 118, 127, 149, 153–56, 160, 161, 174, 178, 180, 182, 184, 185, 188, 248; REDD template, 18, 25, 26, 87, 105, 127, 136, 139, 161, 188, 196–98, 213, 226, 242, 243, 250
transparency, 30, 54, 88, 89, 92–96, 101, 120, 121, 173, 248, 262, 268

uniformity, uniformization, 25, 39, 92, 104, 122, 256, 257, 268
United Nations Framework Convention for Climate Change (UNFCCC), 4–7, 10, 11, 15, 16, 18, 20, 23–28, 30, 65, 68–70, 72–90, 92–105, 109, 110, 113, 114, 116, 118–21, 123–26, 129–31, 135–37, 152, 159, 160, 185, 219, 248, 249, 260–62, 274, 276

vague, language/vagueness, 4, 20, 58, 60, 65, 66, 69, 70, 73, 87, 89, 93–95, 100, 102, 129, 135, 136, 156, 169, 180, 188, 189, 195, 229, 231, 241, 242, 248
value(s), 4, 5, 8, 13, 14, 17, 18, 19, 29, 36, 39, 40, 46, 56, 61, 62, 97, 108, 120, 124, 132, 134, 141, 148, 149–51, 170, 203, 225, 227, 246, 249, 250; exchange value, 50, 53, 55; metrics of, 58, 59, 62; theory of, 49, 54; use value, 53
Verified Carbon Standard (VCS), 140, 185, 190–96, 213, 221, 265, 266, 274

Warsaw framework, 66, 72–74, 79, 81, 84–87, 91–102, 110, 120, 134, 262, 263
World Bank, 18, 20, 24–27, 29, 55, 65, 82, 87, 90, 115, 122, 135, 146, 165, 167, 169, 172–86, 214, 216–26, 228–37, 239–42, 249, 250, 252, 253, 259, 260, 264, 269, 271, 272, 274, 276–78
World Rainforest Movement (WRM), 126, 127
World Wildlife Fund (WWF), 110, 146, 150, 154, 160, 185, 241, 266–68, 272

Xapuri, 28, 145, 146, 199, 201, 203

Yawanawá, 208, 211–13, 274

About the Author

Raquel Rodrigues Machaqueiro holds a PhD in cultural anthropology from George Washington University, where she is currently a postdoctoral associate in the Slave Wrecks' Project and a professorial lecturer. Her research interests include transnational governance and policymaking, development and environmental policies, science and technology studies, and the transatlantic slave trade. She has developed fieldwork in Brazil and in Mozambique. Currently she is working on colonial forestry practices and the financial basis of the Portuguese slave trade.